A MATLAB® Primer for Technical Programming in Materials Science and Engineering

A MATLAB® Primer for Technical Programming in Materials Science and Engineering

Leonid Burstein

Woodhead Publishing is an imprint of Elsevier
The Officers' Mess Business Centre, Royston Road, Duxford, CB22 4QH, United Kingdom
50 Hampshire Street, 5th Floor, Cambridge, MA 02139, United States
The Boulevard, Langford Lane, Kidlington, OX5 1GB, United Kingdom

© 2020 Elsevier Inc. All rights reserved

No part of this publication may be reproduced or transmitted in any form or by any means, electronic or mechanical, including photocopying, recording, or any information storage and retrieval system, without permission in writing from the publisher. Details on how to seek permission, further information about the Publisher's permissions policies and our arrangements with organizations such as the Copyright Clearance Center and the Copyright Licensing Agency, can be found at our website: www.elsevier.com/permissions.

This book and the individual contributions contained in it are protected under copyright by the Publisher (other than as may be noted herein).

Notices

Knowledge and best practice in this field are constantly changing. As new research and experience broaden our understanding, changes in research methods, professional practices, or medical treatment may become necessary.

Practitioners and researchers must always rely on their own experience and knowledge in evaluating and using any information, methods, compounds, or experiments described herein. In using such information or methods they should be mindful of their own safety and the safety of others, including parties for whom they have a professional responsibility.

To the fullest extent of the law, neither the Publisher nor the authors, contributors, or editors, assume any liability for any injury and/or damage to persons or property as a matter of products liability, negligence or otherwise, or from any use or operation of any methods, products, instructions, or ideas contained in the material herein.

Library of Congress Cataloging-in-Publication Data
A catalog record for this book is available from the Library of Congress

British Library Cataloguing-in-Publication Data
A catalogue record for this book is available from the British Library

ISBN: 978-0-12-819115-6 (print)
ISBN: 978-0-12-819116-3 (online)

For information on all Woodhead publications
visit our website at https://www.elsevier.com/books-and-journals

Publisher: Matthew Deans
Acquisition Editor: Glyn Jones
Editorial Project Manager: Emily Thomson
Production Project Manager: Sojan P. Pazhayattil
Cover Designer: Greg Harris

Typeset by SPi Global, India

Dedication

In memory of my parents—Matvey and Leda
To my wife Inna, and my son Dmitri

Contents

Preface ix

1. Introduction   1
   1.1 Some history   1
   1.2 Purpose and audience of the book   2
   1.3 About the book topics   3
   1.4 The structure of the chapters   4
   1.5 About MATLAB® versions   4
   1.6 Order of presentation   4

2. Basics of MATLAB®   7
   2.1 Launching MATLAB   7
   2.2 Vectors, matrices, and arrays   23
   2.3 Flow control   46
   2.4 Questions and exercises for self-testing   59
   2.5 Answers to selected questions and exercises   64

3. Graphics and presentations   67
   3.1 Two- and three-dimensional plots   67
   3.2 Statistical plots   98
   3.3 Supplementary commands for generating 2D and 3D graphs   103
   3.4 Application examples   107
   3.5 Questions and exercises for self-testing   114
   3.6 Answer to selected questions and exercises   118

4. Writing programs for technical computing — 123
- 4.1 Scripts and script files — 123
- 4.2 User-defined functions and function files — 128
- 4.3 Selected MATLAB® functions and its applications in MSE — 132
- 4.4 Live Editor — 144
- 4.5 Application examples — 150
- 4.6 Questions and exercises for self-testing — 161
- 4.7 Answers to selected questions and exercises — 166

5. Curve fitting commands and the Basic Fitting tool — 169
- 5.1 Fitting with polynomials and some other functions — 169
- 5.2 Interactive fitting with the Basic Fitting interface — 176
- 5.3 Single- and multivariate fitting via optimization — 182
- 5.4 Application examples — 185
- 5.5 Questions and exercises for self-testing — 195
- 5.6 Answers to selected questions and exercises — 199

6. ODE-, PDEPE-solvers, and PDE Modeler tool with applications — 205
- 6.1 Ordinary differential equations and ODE solver — 205
- 6.2 Solving partial differential equations with PDE solver — 215
- 6.3 Partial differential equations with the PDE toolbox interface — 221
- 6.4 Application examples — 232
- 6.5 Questions and exercises for self-testing — 251
- 6.6 Answers to selected questions and exercises — 256

Appendix: Characters, operators, and commands for mastering programs   259

Index   265

Preface

Scientists, engineers, and students in materials science and engineering (MSE) perform extensive technical calculations and apply computers and some special programs for this. MATLAB® is a very popular software used in technical computing because of its efficiency and simplicity. This book presents a primer in technical programming in MATLAB with examples from the field of materials science, and addresses a wide MSE audience—undergraduate and graduate students and practicing engineers. It provides the MATLAB fundamentals with a variety of application examples and problems taken from materials engineering, physics of materials and properties of substances, and material phenomenon simulations that should facilitate learning the software language. I hope that many nonprogrammer students, engineers, and scientists from MSE field will find the software convenient for solving their specific problems.

The book accumulates many years of teaching experience in introductory and advanced courses in the fields of MATLAB, material properties, and tribology that were lectured for students, engineers, and scientists specializing in the area in question.

I thank MathWorks Inc.,[a] which kindly granted permission to use certain materials. I am also grateful to the Elsevier Inc.[b] for permission to use the text, tables, figures, and screenshots from my previous book "MATLAB in Quality Assurance Sciences" (Amsterdam—Boston—Cambridge—Heidelberg—London—New York—Oxford—Paris—San Diego—San Francisco—Singapore—Sydney—Tokyo, 2015).

I would also like to express appreciation to Stephen Rifkind, ITA Recognized Translator, who edited the book.

I hope the primer will prove useful to students, engineers, and scientists in MSE areas and enable them to work with the available fine software.

Any reports of errata or bugs, as well as comments and suggestions on the book will be gratefully accepted by the author

Leonid Burstein
ORT BRAUDE COLLEGE OF ENGINEERING, SOFTWARE ENGINEERING, KARMIEL, ISRAEL

---

[a] The MathWorks, Inc., 3 Apple Hill Drive, Natick, MA 01760-2098, United States. Tel: +1-508-647-7000; Fax: +1-508-647-7001. E-mail: info@mathworks.com; Web: www.mathworks.com.

[b] The book was published by the Woodhead Publishing that is an imprint of Elsevier Inc., 50 Hampshire St., 5th Floor, Cambridge, MA 02139, United States. Tel: +1-617-661-7057; Fax: +1-617-661-7061. E-mail: permissions@elsevier.com; Web: https://www.elsevier.com.

Introduction

The world is material: all different natural objects, plants, products, machines, and even ourselves, are designed from various materials. Knowledge of the structure of matter as well as the ability to create new and better materials with the desired properties are among the most important goals of sciences and engineering in general and the science of materials in particular. Naturally, in materials sciences and engineering (MSE), as in other technological areas, calculations and computer modeling are widely used. For this purpose, various kinds of software are used. Among them, MATLAB® has become one of the most widespread and popular in programming for various technical applications. However, a student, teacher, or MSE specialist just beginning to use the MATLAB® soon discovers that each of the available MATLAB® books is designed for a wide range of specialists and that there is no textbook specifically aimed for the problems that are encountered in technical programming in the MSE fields. Overall, a large community of technicians needs a concise, comprehensive text that is easy to understand and provides quick access to the necessary tool. The presented edition aims to fill this gap.

1.1 Some history

The theoretical basics of the MATLAB® language were established in the 1970s by the mathematician Cleve Moler and perfected by the specialists that joined him. First, the language was oriented to the adaptation of the mathematical packages of that time, LINPACK and EISPACK. In a short time, MATLAB® was considered by students and engineers an effective and suitable tool not only for mathematical but also for many technical problems. The language was rewritten in C. Commercial MATLAB® versions have appeared since the mid-1980s in the general software market. By incorporating graphics and development of the special engineering-oriented means—toolboxes—MATLAB® acquired its modern outlines. In general, MATLAB® is a unique assembly of implemented modern numerical methods and specialized tools for engineering calculations developed over the past decades. MATLAB® competes with other software and has established its special place as the software for technical computations. Without going into detail, the following factors and their combinations provide advantages to MATLAB®:

- Versatility and the ability to solve both simple and complex problems with its easy-to-use facilities;
- Highest adaptability to different areas of engineering and science as reflected in a significant number of the problem-oriented toolboxes;

- Convenience and a variety of visualization means for general and specific problems, e.g., MSE problems;
- Quick, simple access to well-organized, extensive documentation.

1.2 Purpose and audience of the book

The purpose of this primer is to provide MSE students, academicians, teachers, engineers, and scientists with a guide that will teach them how to create programs suitable for their professional calculations and present the results in descriptive, graphical, and tabular forms. It is assumed that the reader has no programming experience and will be using the software for the first time. In order to make the primary programming steps and use of commands clear to the target audience, they are demonstrated by problems taken from different areas of materials science whenever possible. Among the variety of available software, MATLAB® distinguishes itself as the tool for technical analysis and calculations. It is renewed and refined in parallel with the developments in modern technology. Modern material sciences specialists intensively used computers with some special programs, and therefore need a universal tool for solving specific problems from their area. Thus, the book serves as a guide to MATLAB® with examples from the field of material engineering and is addressed to undergraduate, graduate, and postgraduate students as well as nonprogrammer technicians that want to master the MATLAB® program to solve problems arising in their areas.

Most of the existing books on the various aspects of MATLAB® can be roughly divided into two kinds: (a) MATLAB® programming books and (b) advanced engineering, science, or mathematics books with MATLAB®-introductory section/s. The first category assumes that the reader is already familiar with math methods and concentrates on programming technique. The second category is generally devoted to special subjects on a somewhat advanced level. This book is different in that it assumes the reader possesses a modest mathematical background and introduces the programming or technical concepts together with a traditional approach. MATLAB® is then used as a tool for subsequent computer solutions, applying it to mechanical and material science problems. An additional distinction of the book is its relatively compact size combined with a variety of examples from a broad range of modern and classical mechanics and material sciences, which help solidify the understanding of the presented material.

In accordance with the foregoing, the principal audiences of the book include:

- Students, engineers, managers, and teachers from the academic and scientific communities in the field of materials science;
- Instructors and their students in MSE study program courses where MATLAB® is used as a supplemental but necessary tool;
- Personnel at materials testing laboratories, students, and nonprogrammers using MATLAB®;

- Students and participants in advanced MSE courses, seminars, or workshops where MATLAB® is taught;
- Scientists who seek to solve MSE-scientific problems and search for similar problems solvable with MATLAB®;
- Self-instructing readers as a means of quick mastering of MATLAB® for their needs.

The book will also serve non-MSE specialists as a reference in numerical applications that require a computer tool for modeling and solving actual engineering problems.

1.3 About the book topics

The topics were selected based on more than 20 years of research experience and 15 years of teaching experience in the fields of tribology, materials and substances properties, and MATLAB®. They were presented so that the beginner can progress gradually using only previously acquired material as prerequisites for each new chapter.

The most important, basic MATLAB® features, including the desktop environment, language design, help options, variables, arithmetical and algebraic functions and operations, matrix and array creation and manipulations with their flow chart control, and conditional statements, are introduced in the second chapter. A command of this material enables the reader to write, execute, and display the simple calculations directly in the command window.

The third chapter presents visualization means by examples of various two- and three-dimensional plots representing the actual calculations. Understanding the material of the second and third chapters allows the reader to create rather complex MATLAB® programs for technical calculations and their graphical representation.

The fourth chapter shows how to create programs in the form of scripts or user-defined functions and then save them as an m-file. The chapter includes the Live Editor, live scripts, and function descriptions and demonstrates their use by the MSE-oriented example. In this chapter, the supplementary commands for common numerical calculations, such as finding the solution of the equation, inter- and extrapolation, differentiation and integration, are discussed together with examples from the MSE fields.

The fifth chapter presents more advanced topics, including the fitting experimental, tabular, or theoretical data. Polynomial fitting, fitting by optimization, and the use of the Basic Fitting interface are described with both single and the multivariate fittings presented.

The final sixth chapter is intended for more advanced readers and explains the specialized commands for solving ordinary differential equations (ODEs), spatially one-dimensional partial differential equations (PDEs), and PDE. The modeler interface for two-dimensional PDEs is briefly presented with examples related to diffusion, heat transfer, and wave equations. To understand this chapter, a familiarity with mathematics on a somewhat higher level is assumed.

The appendix presents a summary collection of approximately 200 studied MATLAB® characters, operators, commands, and functions.

The index contains more than 700 alphabetically arranged names, terms, and commands that were implemented throughout the book.

1.4 The structure of the chapters

Each chapter begins with a general introduction, goals, and chapter content. The new material, main command forms, and its application are then presented. The commands are typically explained in one or two simple forms with possible useful extensions given. Each topic is completely presented in one section in order to allow the readers to attain the knowledge in a focused manner. Tables listing the additionally available commands that correspond to the topic, command description, and examples are included in the chapter.

At the end or middle of the chapter, application problems associated with the MSE area are solved with the commands accessible to the reader. The given solutions are the easiest to understand but not necessarily the shortest or original. Readers would find it useful and are encouraged to try their own solutions and compare the results with those in the book.

For better application, the questions and problems for self-testing are given at the end of each chapter; the first ten questions are MATLAB®-oriented, and all other exercises are related to various MSE disciplines. I recommend solving them to attain better understanding of MATLAB®. At the end of each chapter, the answers to some of the exercises are provided.

It should be noted that the numerical values and contexts used in the various problems of the book are not factual reference data and serve for educational purposes only.

1.5 About MATLAB® versions

Annually, two new versions of MATLAB® appear. Although each version is updated and extended, they allow work with previously written commands. Thus, the basic commands described in this book will remain valid in any future versions. The version used in this book is R2018b (9.5.0.944444). It is expected that readers have MATLAB® installed on their computer and will be able to perform all the basic operations presented in the book.

1.6 Order of presentation

The book presents a MATLAB® primer oriented to a newcomer in computer calculations, with the topics then arranged accordingly. Nevertheless, a teacher is not obligated by this order. For example, the Editor (Section 4.1) can be taught directly after the output commands (Section 2.1.7) to permit the simple program creations on the early learning stages. Likewise, the material regarding creation of script files (Section 4.1) can be studied after the input and output commands (Section 2.2) to allow students to write script programs during the first steps. The polynomial and optimization fitting (Chapter 4) can be presented after the section on 2D graphs (Chapter 2) for better illustration of the fitting results. The ODE and PDE solutions (Chapter 6) can be moved before the fitting (Chapter 4) as their solutions are not connected to curve fitting. Another way to teach MATLAB® is to use the recent Life Editor (described in Chapter 4), which can be introduced to beginners immediately after they become familiar with variables and interactive calculations (prior to Section 2.1.5); this editor can be used throughout the rest of study with examples of scripts/functions converted to a live file (as explained in Section 4.4.3).

In general, I hope the book will help MATLAB® users study the software and apply it to MSE problems.

2

Basics of MATLAB®

Almost half a century has passed since a special computer tool was created for mathematicians. The tool and its language were named MATLAB®. The term is derived from the three first letters of two words, *matrix* and *laboratory*, to emphasize the main element of this language: a matrix. This matrix-based approach unifies the calculating procedures for both algebraic and graphical processing. Within a short period, MATLAB was adapted and has become a convenient tool for technical computing. Today, engineers use it in a variety of applied and scientific calculations of various fields, including material engineering. Over time, MATLAB has undergone significant changes. It has become more diverse, effective, and complex. Therefore, it is necessary to study the available tool with the goal of efficient technical programming for solving actual problems.

In this chapter, the tool desktop, its toolstrip, and available windows are presented; the starting procedure is introduced; the main commands for simple arithmetical, algebraic, matrix, and array operations are described; and finally, the basic loops and relational and logical operators are explained.

2.1 Launching MATLAB

To get started with MATLAB, the user should run the software. It is assumed that this software was previously installed on your computer. MATLAB and other application programs, as well as all computer resources, are controlled by a special set of programs called the operating system (OS). These systems may vary with the computing platform. Thus, we additionally assume here that the MATLAB is managed by the Windows OS. To launch MATLAB, click on the icon  showing a red L-shaped membrane. This icon is usually placed in the Windows Start Menu, on the Desktop, and on the Task Bar (Fig. 2.1). Alternately, it is possible to start MATLAB by printing the word MATLAB in the Windows Search box. You may also start the program by finding and clicking on the matlab.exe file in the MATLAB bin-directory with the path to this file being C:\ProgramFiles\MATLAB\R2018a\bin\matlab.exe. The primary information appears in the startup nameplate with the image of the MATLAB logo and some product information: the opening version and license numbers and the developer company name. The nameplate appears for a short time and disappears before opening the MATLAB Desktop.

2.1.1 MATLAB Desktop and its toolstrip and windows

The view of the resulting MATLAB Desktop is shown in the Fig. 2.2. The desktop contains the toolstrip and three windows: Command, Current Folder, and Workspace.

8 A MATLAB® Primer for Technical Programming in Materials Science and Engineering

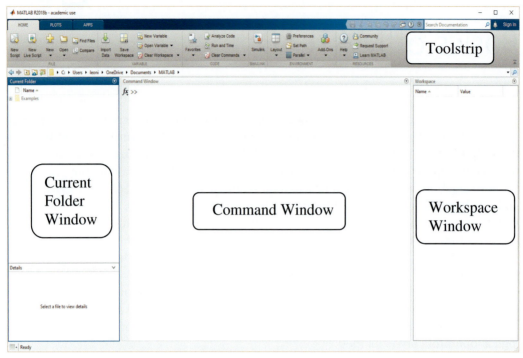

FIG. 2.1 MATLAB R2018b Desktop, default view.

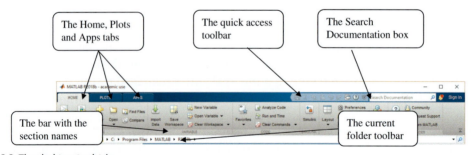

FIG. 2.2 The desktop toolstrip.

2.1.1.1 Toolstrip

The toolstrip contains the main MATLAB purpose operations and functions, presented in three global tabs: the Home, Plots, and Apps. The tabs are divided into sections that contain a series of related controls, including buttons, drop-down menus, and other user interface elements (Fig. 2.2). The tabs include a number of buttons that are grouped in sections according to functionality (file, variable, code, etc.). The sections include buttons for the execution of various operations (e.g., Open, New Variable, Preferences, etc.). The

Home tab is the tab used most intensively and includes general purpose operations, such as creating new files, importing data, and managing your workspace as well as the settings of the Desktop layout. The Plots tab displays a gallery of plots available in MATLAB and any toolboxes that you have installed. The Apps tab contains a gallery of applications from the toolboxes.

The quick access toolbar is located to the right of the top of the strip. This toolbar contains frequently used options such as Cut, Copy, and Paste. The Search Documentation box is located next to the quick access toolbar and enables you to search for documentation about commands and other topics of interest.

The line with the current folder toolbar is placed on the bottom of the strip. This toolbar controls the currently working directory.

2.1.1.2 Command window

This window is the main component of the desktop where the commands are entered and the result of their execution is displayed. For convenience, the window can be separated from the desktop by choosing the ⬈ Undock line of the Show Command Window Actions button ⊙ to the right of the window title bar. It is possible to separate all desktop windows. To assemble separated windows, choose the ⬋ Dock line or select the Default line in the Desktop Layout of the Desktop menu section.

2.1.1.3 Workspace window

In this window, the variable icons and other objects currently located in the MATLAB workspace appear. It also displays the class of each variable and its values. These data are automatically updated with calculations.

2.1.1.4 Current folder window

This browser window shows the full path and contents of the current folder. When starting MATLAB here, we can view the starting directory, called the startup directory, as well as the files and folders located there. The window has the Details panel located at its bottom where information about the selected file appears.

The above windows are used most intensively. Other windows include Command History, Help, Editor, Live Script Editor, and Figure, which do not appear by default when opening the desktop. They are described within the chapters where they are used.

2.1.2 Interactive calculations with elementary math functions

MATLAB commands can be entered and executed in two ways: interactively in the Command Window directly and by a program containing the commands previously written and saved in the m-files. The interactive way is briefly presented here while the second will be explained later.

To enter and realize a command, it must be typed in the Command Window immediately after the command prompt >>, which indicates the place to type the command. Fig. 2.3 presents some elementary commands entered in the Command Window.

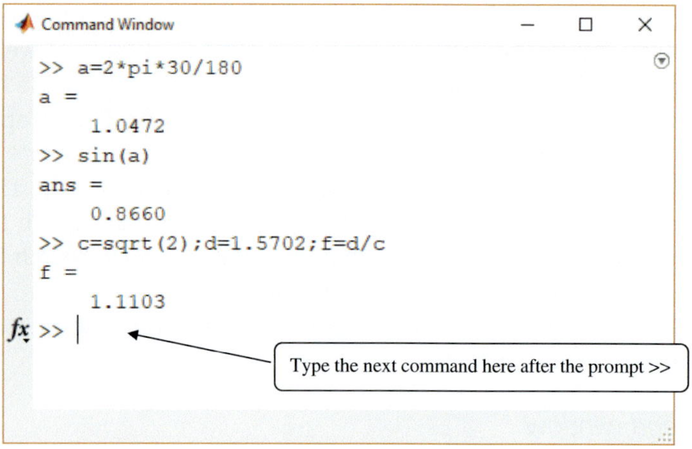

```
>> a=2*pi*30/180
a =
    1.0472
>> sin(a)
ans =
    0.8660
>> c=sqrt(2);d=1.5702;f=d/c
f =
    1.1103
fx >> |
```

Type the next command here after the prompt >>

FIG. 2.3 Undocked Command Window with some elementary commands.

The function button *fx*, appearing on the empty command line to the left of the prompt, is called the Function browser. It helps us find the needed commands and information about their syntaxes and usage.

Entering a command and manipulating it require us to master the following rules:

- The command must be typed next to the prompt >>;
- To execute the command, the Enter key **Enter** must be pressed;
- A command in a preceding line cannot be changed; to correct or repeat an executed command, press the up-arrow key ↑, and then introduce the desirable change in the shown command followed by pressing the Enter key;
- A long command can be continued on the next line by typing ellipsis ... (three periods); the total length of a single command may not exceed 4096 characters;
- Two or more commands can be written in the same line; in this case, they should be divided by semicolons (;) or by commas (,);
- A semicolon entered at the end of the command withholds the display of the answer;
- The symbol % (percent symbol) designates the comments that must be written after it in the line; the comments are not executed after entering;
- To clear the Command Window, enter the `clc` command from the command line;
- The result of entering a command is a variable with the name `ans`;
- The equal sign (=) is called the assignment operator and is used to specify a value for a variable to the `ans`;
- An entered/computed new value cancels its predecessor in the variable, e.g., if the value of *a* was assigned to be 4.3118, after entering the new command, >> a=16.3214, the previous value of *a* is removed with *a* receiving the 16.3214 value.

The Command Window can be used as a calculator by using the following symbols for arithmetical operations: + (addition), - (subtraction), * (multiplication), / (right division), \ (left division, used mostly for matrices), ^ (exponentiation).

The given symbols are used in arithmetic, algebraic, or other math calculations, which manipulate using a wide variety of elementary, trigonometric, and specific mathematical functions. The MATLAB commands for calculations with these functions must be written as the name with the argument in parentheses, e.g., cos x, be written as cos(x). Argument x is given in radians in all trigonometric functions. If the appropriate command is written with the ending letter d, e.g., cosd(x), the argument x should be given in degrees. The inverse trigonometric functions with the ending d produce the result in degrees. A short list of commands for calculations with math functions is given in Table 2.1. Hereinafter, the commands/operators/variables that are entered in the Command Window are written

Table 2.1 Some commands for computing the elementary, trigonometric, and specific mathematical functions (alphabetically).

MATLAB command	Notation	Example of usage (input and output)
abs(x)	$\|x\|$—absolute value	>> abs(-17.1243) ans = 17.1243
acos(x)	arccos x—inverse cosine	>> acos(0.5) ans = 1.0472
acosd(x)	arccosd x—inverse cosine with x between −1 and 1; result in degrees between 0 and 180 degrees	>> acosd(0.5) ans = 60.0000
acot(x)	arccot x—inverse cotangent	>> acot(0.5) ans = 1.1071
acotd(x)	arccotd x—inverse cotangent; result in degrees between −90 and 90 degrees	>> acotd(0.5) ans = 63.4349
asin(x)	arcsin x—inverse sine	>> asin(0.5) ans = 0.5236
asind(x)	arcsind x – inverse sine with x between −1 and 1; result in degrees between −90 and 90 degrees	>> asind(0.5) ans = 30.0000
atan(x)	arctan x—inverse tangent	>> atan(0.5) ans = 0.4636
atand(x)	arctand x—inverse tangent; result in degrees between −90 and 90 degrees (asymptotically)	>> atand(0.5) ans = 26.5651
ceil(x)	Round towards infinity	>> ceil(2.3) ans = 3

Continued

Table 2.1 Some commands for computing the elementary, trigonometric, and specific mathematical functions (alphabetically)—cont'd

MATLAB command	Notation	Example of usage (input and output)
cos(x)	cos x—cosine	>> cos(pi/6) ans = 0.8660
cosd(x)	cosd x—cosine with x in degrees	>> cosd(30) ans = 0.8660
cosh(x)	Cosh x—hyperbolic cosine	>>cosh(pi/6) ans = 1.1402
cot(x)	cot x—cotangent	>> cot(pi/6) ans = 1.7321
cotd(x)	cotd x—cotangent with x in degrees	>> cotd(30) ans = 1.7321
coth(x)	coth—hyperbolic tangent	>> coth(pi/6) ans = 2.0813
erf(x)	$\mathrm{erf}(x) = \frac{2}{\sqrt{\pi}} \int_0^x e^{-t^2} dt$—error function	>> erf(2/sqrt(3)) ans = 0.8975
exp(x)	e^x—exponential function	>>exp(3.1) ans = 22.1980
factorial(n)	n!—factorial; product of the integers from 1 to n	>> factorial(6) ans = 720
fix(x)	Round toward zero	>> fix(-2.6) ans = -2
floor(x)	Round towards minus infinity	>> floor(-2.3) ans = -3
gamma(x)	$\Gamma(x) = \int_0^\infty e^{-t} t^{x-1} dt$—gamma function	>> gamma(4.9) ans = 20.6674
log(x)	ln x—natural (base e) logarithm	>> log(2) ans = 0.6931
log10(x)	$\log_{10} x$—Napierian (base 10) logarithm	>> log10(2) ans = 0.3010
log10(x)/log10(a)	$\log_a x$—base a logarithm	>> log10(2)/log10(2.7183)

Table 2.1 Some commands for computing the elementary, trigonometric, and specific mathematical functions (alphabetically)—cont'd

MATLAB command	Notation	Example of usage (input and output)
pi	π —the number π (circumference-to-diameter ratio of circle)	ans = 0.6931 >> 3/2*pi ans = 4.7124
round(x,n)	Round to the nearest decimal or integer	>>round(2.7865,2) ans = 2.7900
sin(x)	sin x—sine	>> sin(pi/6) ans = 0.5000
sind(x)	sind x—sine with x in degrees	>> sind(30) ans = 0.5000
sinh(x)	sinh—hyperbolic sine	>> sinh(pi/6) ans = 0.5479
sqrt(x)	$\sqrt{x}$—square root	>> sqrt(3/5) ans = 0.7746
tan(x)	tan x—tangent	>> tan(pi/6) ans = 0.5774
tanh(x)	tanh x—hyperbolic tangent	>> tanh(pi/6) ans = 0.4805
tand(x)	tand x—tangent with x in degrees	>> tand(30) ans = 0.5774

Based on Burstein, L. (2015). *Matlab in Quality Assurance Sciences*. Cambridge: Elsevier-WP.

after the command line prompt (>>). For further command execution, the user must press the Enter key after entering one or more commands written in one command line.

Arithmetic operations are performed in the following order: operations in parentheses (starting with the innermost), exponentiation, multiplication and division, addition and subtraction. If the expression contains operations of the same priority, they run from left to right.

Examples of arithmetic operations in the Command Window with explanations about the order of perform are given below:

The resulted numbers are shown on the display in the short format (default format)—a fixed point followed by four decimal digits, with the last decimal digit rounded. The format of the displayed number can be changed to long, specifically 14 digits after the point,

14   A MATLAB® Primer for Technical Programming in Materials Science and Engineering

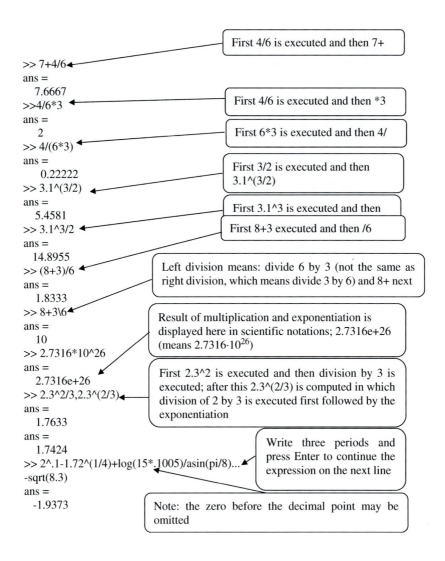

by typing the command: >> `format long`. To return to the default format, type `format` (additional information on formats is provided in Section 2.1.6).

2.1.3 Help commands and Help Window

To obtain information regarding the command functionality, type and enter `help` and the command name after a space. For example, entering the `help format` command provides explanations about available output formats and gives examples of its usage strictly in the Command Window. For a command concerning a topic of interest, the `lookfor` command may be used. For example, for the name of MATLAB command/s on the subject of the density, you can enter `lookfor density`. After a rather long search, the commands will appear on the screen, as shown below (incomplete).

Chapter 2 • Basics of MATLAB® 15

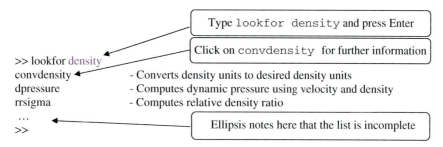

For further information, click on the selected command or use the `help` command. To interrupt the search process, the two abort keys `ctrl` and `c` should be clicked together. These keys may also be used to interrupt any other process, including execution of a program/command.

Note:

The information returned by the `lookfor` command can vary on different computers as this is determined by the toolbox set installed with MATLAB. For example, for the described request about the `convdensity` command, the Aerospace toolbox™ must be installed on your computer.

For more detailed information, the `doc` command can be used, i.e., `doc convdensity`. In this case, the Help window will open (Fig. 2.4).

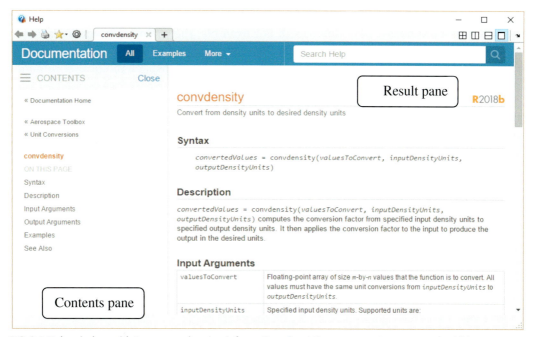

FIG. 2.4 Help window with two panes showing information about the `convdensity` command, which converts density units.

This window can be opened in other ways, such as by selecting the Documentation line in the popup menu of the Help button on the toolstrip of the Resources group of the desktop Home tab.

When the Help window is opened with the `doc` command, the Help window comprises the menu line, the Search Documentation field, the narrow Contents pane (to the left) with Contents button and the broad pane with documentation chapters containing the searched topic, as well as the search results with the information on the subject. Information can also be obtained by typing the word/s into the Search Documentation field at the top of the window. In this case, the Contents pane represents the defined information by product, category, or type. In addition, the Results pane presents a preview of the defined information and toolbox name for which the information is relevant. For example, the `Documentation > MATLAB > Language Fundamentals > Matrices and Arrays` line designates the location of the page with explanations on the requested command/example/code.

2.1.4 About toolboxes

Although elementary and specific mathematical functions, e.g., `sin`, `cos`, `sqrt`, `exp`, and `log`, are used in a wide range of sciences—from aeronautics to medicine. In each area, there are specific problems that require some special commands for their solutions. For these purposes, basic and several problem-oriented tools and concentrated commands in the so-termed toolboxes, were developed. For example, the main commands discussed are collected in the MATLAB toolbox, with the commands related to statistics in the Statistics toolbar, commands for aerospace problems in the Aerospace toolbox, commands for neural networks in the Neural Network toolbox, etc.

To verify which toolboxes are available on your computer, the `ver` command is used. This command, when typed and entered in the Command Window, displays the header with product information and a list of toolbox names, versions, and releases as follows:

```
>> ver
```

```
MATLAB Version: 9.5.0.944444 (R2018b)
MATLAB License Number: XXXXXXXX
Operating System: Microsoft Windows 10 Pro Version 10.0 (Build 17134)
Java Version: Java 1.8.0_152-b16 with Oracle Corporation Java HotSpot(TM) 64-Bit Server VM mixed mode
```

```
MATLAB                              Version 9.5      (R2018b)
Simulink                            Version 9.2      (R2018b)
Aerospace Blockset                  Version 4.0      (R2018b)
```

```
Aerospace Toolbox                    Version 3.0    (R2018b)
Antenna Toolbox                      Version 3.2    (R2018b)
Audio System Toolbox                 Version 1.5    (R2018b)
Automated Driving System Toolbox
...
```

The represented toolbox list was interrupted as it can be quite long, depending on the installed toolboxes. The information about available toolboxes also can be obtained by selecting the Manage Add-Ons option in the Add-Ons button of the Home tab on the MATLAB toolstrip.

2.1.5 About variables and commands to variable management

In programming, a variable is a symbol, namely letter/s and number/s to which some specific numerical value/s is/are assigned. It is possible to assign a single number (a scalar) or a table of numbers (an array) to a variable. MATLAB allocates space in the computer's memory for storage of both the variable names and their values. The variable name can contain as many as 63 characters long and consists of letters, digits, and underscores, the first character being a letter. Existing command names (sin, cos, sqrt, etc.) and arithmetic operations (+, −, *, /, \) are not recommended for use as variable names because they would confuse the system.

The assignment and use of variables in algebraic calculations are demonstrated in the following screenshot:

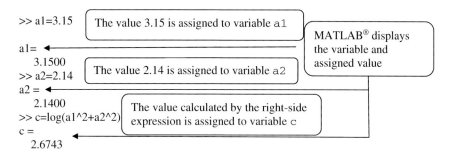

Specific variables/constants are assigned and permanently stored in MATLAB. Such variables are termed *predefined*. In addition to previously mentioned pi and ans, other predefined variables, include inf (infinity, the result, for example, of division by zero), i or j (square root of −1), NaN (not-a-number, used when a numerical value is moot, e.g., 0/0) and eps (smallest allowed difference between two numbers).

The following commands manage the variables: clear—for removing all variables from memory, or clear var1 var2 ...—for removing named variables only; who—for displaying the names of variables and whos—for displaying variable names, matrix sizes, variable byte sizes, and variable storage classes (numeric variables are double precision by default). In addition, each variable with the same information, as in the case of whos, and those with additional data are presented in the Workspace Window (marked by the icon ▦; this icon varies for different variable classes). To obtain additional information about a variable, click the mouse's right button with the cursor placed on the variable header line. The pop-up menu appears with a list of possible additional information.

2.1.6 Output formats

MATLAB displays the result of a command execution on the screen in a specific format, which is governed by the `format` command. The command takes the form

`format` or `format` `format_type`

The first command sets a short format type, which is the default format of numeric data. In this format, the four decimal digits are displayed, e.g., 3.1416, with the last digit rounded. When the real number is lesser than 0.001 or greater than 1000, the number is shown in the `shortE` format. The latter format uses scientific notations—a number between 1 and 10 multiplied by a power of 10 (e.g., the Plank's constant in scientific notations is presented as 6.6261e-34, in m kg/s and should be read as $6.6261 \cdot 10^{-34}$ while the rubber Young's modulus is presented as 14.5000e-3 and should be read as $14.5 \cdot 10^{-3}$). The number $a=1000.1$ is displayed in the same in `short` and `shortE` formats as 1.0001e+03 where e+03 is 10^3 while the whole number should be read as $1.0001 \cdot 10^3$. Note, scientific notations also can be used for variable input, e.g., `d=3.4e6` inputs value $3.4 \cdot 10^6$ in the `d` variable.

The *format_type* parameter in the second option of the presented `format` commands is a word that specifies the type of the displayed numbers. There are more than two described format types. Thus, to show 15 decimal digits, the `long` or `longE` format should be used. For example, when setting the longE format type and inputting the Boltzmann constant $1.38064852 \cdot 10^{-23}$, $m^2 kg/s^2/K$, MATLAB yields the following results:

```
>> format longE
>> B=1.38064852e-23
B =
    1.380648520000000e-23
```

Note:

- The `format` commands change the length of the numbers on the display but do not make changes in the computer memory nor do they affect the type of the inputted numbers.
- Once a certain number format type is specified, all subsequent numbers are displayed within it.
- To return to the replaced `short` format, the `format` or `format short` command must be entered.

- The *format_type* parameter can be written with a space between short and E and long and E while the capital E can be written as a conventional e.

The short or long formats are written with the G letter display, respectively, the best 5-digit or 15-digit number. For example, in the longG format, the Boltzmann constant appears without trailing zeros:

```
>> format longG
B =
        1.38064852e-23
```

This format is ideal for cases that display matrices or arrays with numbers very different in size since the format picks where usual (e.g., short) and scientific (e.g., shortE) notation is best used.

More details about the described formats and others can be obtained by using the help format or doc format command.

2.1.7 Output commands

MATLAB displays the result immediately after entering a command (but not when a semicolon follows the command) accordingly to assigned format. For additional output management, the two most frequently used commands are disp and fprintf.

The first of these commands displays texts or variable values without the name of the variable and the = (equal) sign. Every new disp command yields its result in a new line. In general form, the command reads

```
disp('Text string')  or  disp(Variable name)
```

The text between the quotes is displayed in blue.
For example:

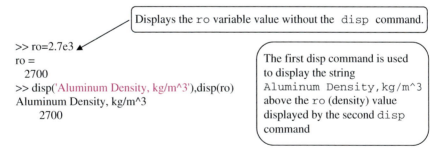

The second command, fprintf, is used to display texts and data or save them in a file. For displaying calculation results, the various forms of this command can be used. Most of them present difficulties for beginners. So, here we give the simplest forms of the command.

For formatted output and displaying text and a number on the same line, the following form is used:

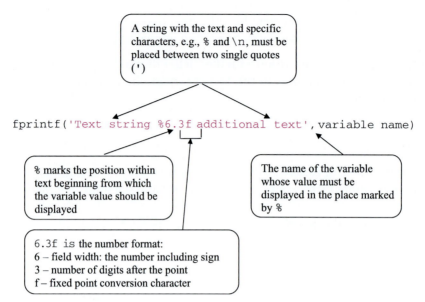

To display text with a new line or divide text on two or more lines, write the \n (slash n) before the word that you want to see on the new line. The same characters should be written to cause the appearance of the >> prompt from the new line after executing the `fprintf` command. The field width number and number of digits after the point (6.3 in the presented example) are optional but the % sign and the character f, referred to as conversion characters, are obligatory. In other words, if %f was written instead of %6.3f, the number will be displayed with six digits after the point. The f specifies the fixed point notation in which the number is displayed and leads to a real number conversion to the fixed number of digits after the decimal point. Some additional notations that can be used include i (or d)—integer, e—exponential, e.g., 2.309123e+001, and g—the more compact form of e or f, with no trailing zeros. The addition of several %f units (or full formatting elements) permits the inclusion of multiple variable values in the text.

The following example uses the `fprintf` command:

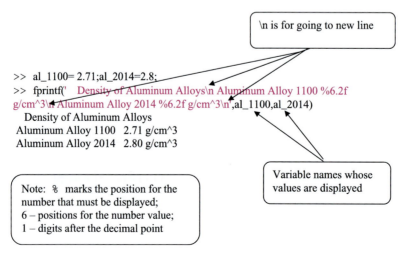

As in `disp`, the color of the characters in the quotes is blue.

The presented output commands also can be used to display tables. This will be shown later after studying the vectors, matrices, and arrays (see Section 2.2.5).

2.1.8 Application examples

Some examples of simple calculations by mathematical expressions are provided below.

2.1.8.1 Bravais lattice: Volume of the triclinic crystal unit cell

A material has the unit cell of the triclinic crystal form with the following lattice parameters: $a=4.05$ Å, $b=4.85$ Å, $c=2.54$ Å, $\alpha=\beta=\pi/2$, and $\gamma=1.8326$ rad. The volume of the unit cell can be calculated with the expression:

$$V = abc\sqrt{1- \cos^2\alpha - \cos^2\beta - \cos^2\gamma + 2\cos\alpha\cos\beta\cos\gamma}$$

Problem: Calculate the unit cell volume with the given expression. Display the result with and without the `disp` command.

The solution:

```
>> a=4.05; b=4.85; c=2.54;
>> alpha=pi/2; beta= alpha; gamma=1.8326;
>> V=a*b*c*sqrt(1-cos(alpha)^2-cos(beta)^2-...
cos(gamma)^2+2*cos(alpha)*cos(beta)*cos(gamma))
V =
   48.1919
>> disp('Volume, Angstroms^3'),disp(V)
Volume, Angstroms^3
   48.1919
```

- Assigns values to the a, b, c, alpha, beta and gamma variables
- Calculates the cell volume; three points designate that a part of command moves to the next line
- Displays the volume using the `disp` command

2.1.8.2 Brinell hardness number

Hardness number in Brinell scale, BHN, is calculated by the expression

$$BHN = \frac{2P}{\pi D\left(D-\sqrt{D^2-d^2}\right)}$$

where applied load P, steel boll diameter D and indentation diameter d are defined in a penetrating test.

Problem: Calculate the BHN for a specimen with $P=3000$ kg, $D=10$ mm, and $d=5$ mm. Display the result with and without the `disp` command.

```
>> P=3000; D=10; d=5;
```
Assigns values to the P, D, and d variables

```
>> BHN=2/(pi*D*(D-sqrt(D^2-d^2)))
BHN =
    0.0475
```
Calculates the BHN value.

```
>> disp('Brinell Hardness,kg/mm^2'),disp(BHN)
Brinell Hardness,kg/mm^2
    0.0475
```
Displays BHN using the `disp` *command*

2.1.8.3 Friction coefficient

To move a block with mass *m*, force *F* are required. The coefficient of friction, μ, can be defined with the equation

$$\mu = \frac{F}{mg}$$

Problem: Calculate the friction coefficient with $F = 12.8\,\text{N}$, $m = 2.1\,\text{kg}$, and $g = 9.81\,\text{m/s}^2$.

```
>> F=12.8;m=2.1;
>> g=9.81;
>>mu=F/(m*g)
mu =
    0.6213
```
Assigns values to the F, m, and g

Calculates the friction coefficient

2.1.8.4 Stress intensity factor

For a finite plate, the stress intensity factor of the centrally located crack can be calculated by the expression:

$$K = \sigma\sqrt{\pi a}\left[\frac{1 - \frac{a}{2b} + 0.326\left(\frac{a}{b}\right)^2}{\sqrt{1 - \frac{a}{b}}}\right]$$

where *a* is the crack radii, *b* is the half width of the plate, and σ is the uniaxial tension.

Problem: Calculate stress intensity factor K, $\text{kPa}\cdot\text{m}^{1/2}$ when $a = 0.0381\,\text{m}$, $b = 0.1016\,\text{m}$, and $\sigma = 82.7371\cdot 10^3\,\text{kPa}$.

The solution:

```
>> a=.0381;b=.1016;
>> sigma=82.7371e3;
>> ab=a/b;
>> K=sigma*sqrt(pi*a)*(1-ab/2+0.326*ab^2)/sqrt(1-ab)
K =
   3.1078e+04
```

2.2 Vectors, matrices, and arrays

A normal table of numbers, which often represents data in materials science, is an array or matrix. The variables that were used heretofore were given in scalar form. Each of these variables is a 1×1 matrix in MATLAB. Mathematical operations with matrices and arrays are more complicated than with scalars: linear algebra operations should be applied for matrices while element-wise operations for arrays.

2.2.1 Vectors and matrices: Generation and handling

2.2.1.1 Generation of vectors

Vectors are sequentially written numbers organized in rows or in columns and termed, respectively, row or column vectors. They also can be presented as row of words or equations.

In MATLAB, a vector is generated by typing the numbers in square brackets. In case of row vector, the spaces or commas must be typed between the numbers. In the case column vector, the semicolons are typed between the numbers. Pressing Enter between the numbers also generates a column vector.

An example is the data of elasticity modulus of a material measured at different temperatures shown in Table 2.2.

This data can be presented as two vectors:

```
>> temperature=[0 83.3 166.5 250 333 416.5 450]
temperature =
     0   83.3000  166.5000  250.0000  333.0000  416.5000  450.0000
>> elasticity=[210;200;190;175;160;145;103]
elasticity =
   210
   200
   190
   175
   160
   145
   103
```

Numerical data from the first row of Table 2.2 is assigned to the row vector entitled `temperature`

Numerical data from the second row of Table 2.2 is assigned to the column vector entitled `elasticity`

Table 2.2 Elasticity modulus-temperature data.

Temperature (°C)	0	83.3	166.5	250	333	416.5	450
Elasticity modulus (GPa)	210	200	190	175	160	145	134

If there is not enough space to display all the vector elements in one line, they are displayed in two or more lines with a message informing which column is presented in each line. For example:

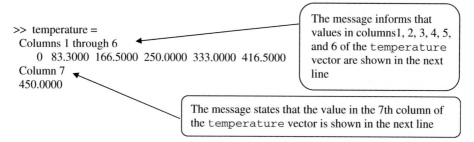

Vectors also can represent the position of a point in three-dimensional coordinates, such as point A in Cartesian coordinates, that are presented by a position vector $r_A = 3i + 5j + 2k$ (where i, j, and k are unit vectors and 2, 4, and 5 are the vector projections on the axis), which can be written as row vector A=[3,5,2].

Frequently, numbers of a vector differ by the same value. For example, in vector v=[1 3 5 7 9], the spacing between the elements is 2. To create such vectors, there are two operators namely: ':' (colon) and `linspace`.

The `colon` operator has the form

```
vector_name=a:q:b
```

where `a` and `b` are, respectively, the first and last term in the vector, while `q` is the step between adjacent terms within it. The last number in generated vector cannot exceed the last term `b`. The step `q` can be omitted. In this case, the step is set to 1 by default. Examples are:

```
>> pressure=1:0.012:1.1
pressure =
  Columns 1 through 6
    1.0000  1.0120  1.0240  1.0360  1.0480  1.0600
  Columns 7 through 9
    1.0720  1.0840  1.0960
```
First number 1, last number does not exceed 1.1, step 0.012

```
>> x=-4:4
x =
  -4  -3  -2  -1  0  1  2  3
```
First number -4, last number 4; step by default is 1

```
>> y=0.3:10.9/10.13:1
y =
    0.2000  0.3200  0.4400  0.5600  0.6800  0.8000  0.9200
```
First number 0.3 while the last number does not exceed 1, a step defined by division 10.9/10.13 (= 1.0760)

```
>> z=13.2:-2.21:1.33
z =
   13.2000  10.9900  8.7800  6.5700  4.3600  2.1500
```
The first number 13.2 with the last number is not less than 1.33, step -2.21

The `linspace` operator has the form:

`vector_name=linspace(a,b,n)`

where `a` and `b` are the first and last numbers, respectively, and `n` is the amount of equally spaced numbers. When `n` is not specified, this value equals 100 by default. For example:

```
>> x=linspace(0,28,8)
x =
    0    4    8   12   16   20   24   28
```
8 equally spaced numbers, first number 0, last number 28

```
>> y=linspace(-10,100,3)
y =
   -10   45  100
```
3 equally spaced numbers, first number -10, last number 100

```
>> z=linspace(15.2,1.3,5)
z =
   15.2000  11.7250   8.2500   4.7750   1.3000
```
5 equally spaced numbers, first number 15.2, last number 1.3

```
>> v=linspace(0,100)
v =
  Columns 1 through 8
        0   1.0101   2.0202   3.0303   4.0404   5.0505   6.0606   7.0707
...
```
The amount of numbers is omitted, with default being 100, first number 0, last number 100. Only the first 8 values are shown

The position of an element in the generated vector is its address, which should be a positive integer and not zero, as in the following examples:

- The fifth position in the eighth element vector `elasticity` above can be addressed as `elasticity (5)`; the element located here is 116.
- If the `elasticity (0)` command was entered, an error message appears since the element number cannot be zero.

The last position in a vector may be addressed with the `end` terminator, e.g., `elasticity (end)` is the last position in the `elasticity` vector and marks the number located here, 121. Another way to address the last element is to give the position of number, namely `elasticity (8)`.

2.2.1.2 Generation of matrices and arrays

An array or two-dimensional matrix has rows and columns with numeric data and resembles a numeric table. The difference between a matrix and the ordinal table is manifested only when certain mathematical operations are performed on the matrices. When the number of rows and columns is equal, the matrix is called square. Otherwise, it is rectangular. Like a vector, a matrix is generated by entering the elements in square brackets with spaces or commas between them and with a semicolon between the rows

Table 2.3 Air compressibility factor Z (shaded).

Temperature (K)	Pressure (bar)		
	1	10	20
300	0.9999	0.9974	0.9950
400	1.0002	1.0025	1.0046
500	1.0003	1.0034	1.0074
600	1.0004	1.0039	1.0081

or by pressing Enter between the rows. The number of elements in each row must be equal. The elements also can be given with variable names or mathematical expressions.

As an example, Table 2.3 presents air compressibility factor Z (dimensionless) performed for three different pressures (in bar) and four different temperatures (in K):

The matrix presentation is displayed in the shaded parts of the table, and some other matrix generation examples are:

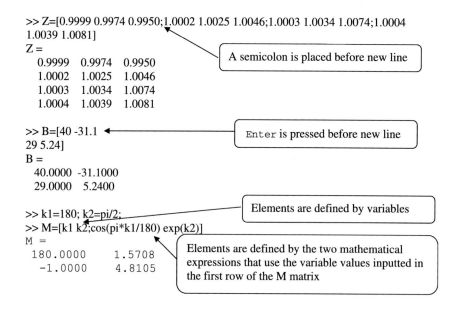

To refer to matrix element/s, row-column addressing is used. For instance, in matrix Z in the previous example, set Z(2,3) refers to the number 1.0046 and Z(3,2)—the number 1.0034. Row or column numbering begins with 1. Therefore, the first element in matrix Z is Z(1,1).

For sequential elements or an entire row or column, the semicolon can be used. For example, Z(2:3,2) refers to the numbers in the second and third rows of the column 2 in matrix Z while Z(:,n) refers to the elements of all rows in column *n* and Z(m,:) to those of all the columns in row *m*.

In addition to row-column addressing, linear addressing can be used. In this case, a single number is used instead of two (row and column) numbers. The elements place within the matrix is indicated sequentially, beginning with the first element of and along the first column, then continuing along the second column and so forth, up to the last element in the last column. For example, for Z matrix the Z(6) command refers to element Z(2,2), Z(8) to the Z(4,2) while Z(5:8) is the same as Z(:,2), etc.

It is possible also to generate a new matrix by combining an existing matrix with a vector or with another matrix using the square brackets. Examples of this kind are presented below using Z matrix (previous example) and the pressure row and temperature column from Table 2.3.

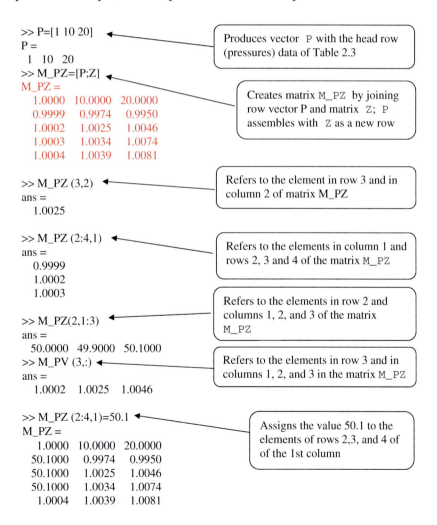

To convert a row/column vector into a column/row one and for the rows/columns exchange in matrices, the `transpose` command or sign ' (quote) is applied, as illustrated:

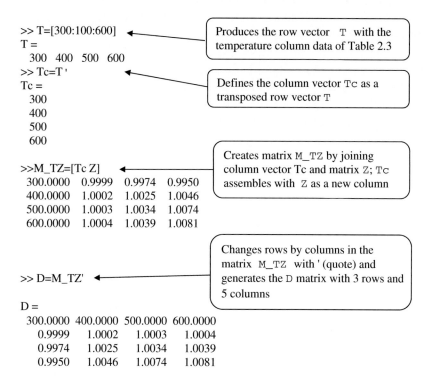

2.2.2 Matrix operations

When computing, vectors, matrices, and arrays are used in various mathematical operations similar to operations with single variables as illustrated below.

2.2.2.1 Addition and subtraction

Two basic arithmetical operations—addition and subtraction—are performed element by element in the matrices that must be equal in size, e.g., when *M1* and *M2* are two matrices sized 3×2 each:

$$M1 = \begin{bmatrix} M1_{11} & M1_{12} \\ M1_{21} & M1_{22} \\ M1_{31} & M1_{32} \end{bmatrix} \text{ and } M2 = \begin{bmatrix} M2_{11} & M2_{12} \\ M2_{21} & M2_{22} \\ M2_{31} & M2_{32} \end{bmatrix}$$

the sum of these matrices is

$$M1 + M2 = \begin{bmatrix} M1_{11} + M2_{11} & M1_{12} + M2_{12} \\ M1_{21} + M2_{21} & M1_{22} + M2_{22} \\ M1_{31} + M2_{31} & M1_{32} + M2_{32} \end{bmatrix}$$

In addition and subtraction operations, the commutative law is valid, namely $M1 + M2 = M2 + M1$.

2.2.2.2 Multiplication

Multiplication of matrices is a more complicated operation. In accordance with the rules of linear algebra, this is feasible only when the number of elements of a row in the first matrix equals that of the column elements in the second matrix, i.e., the inner dimensions of the matrices must be equal. Thus, the above matrices $M1$, 3×2, and $M2$, 3×2, cannot be multiplied unless $M2$ is replaced by another, sized 2×3,

$$M2 = \begin{bmatrix} M2_{11} & M2_{12} & M2_{13} \\ M2_{21} & M2_{22} & M2_{23} \end{bmatrix}$$

the inner dimensions of the matrices $M1$ and $M2$ are equal and multiplication becomes possible.

$$M1*M2 = \begin{bmatrix} M1_{11}M2_{11} + M1_{12}M2_{21} & M1_{11}M2_{12} + M1_{12}M2_{22} & M1_{11}M2_{13} + M1_{12}M2_{23} \\ M1_{21}M2_{11} + M1_{22}M2_{21} & M1_{21}M2_{12} + M1_{22}M2_{22} & M1_{21}M2_{13} + M1_{22}M2_{23} \\ M1_{31}M2_{11} + M1_{32}M2_{21} & M1_{31}M2_{12} + M1_{32}M2_{22} & M1_{31}M2_{13} + M1_{32}M2_{23} \end{bmatrix}$$

It is not difficult to verify that the product $M1*M2$ is not the same as $M2*M1$. In other words, the commutative law does not apply to multiplication of the matrices.

Various examples of matrix addition, subtraction, and multiplication are given below.

30   A MATLAB® Primer for Technical Programming in Materials Science and Engineering

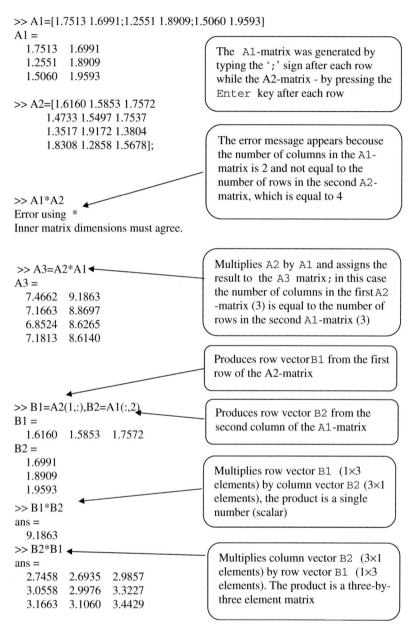

A convenient application of matrix multiplication in technical programming is the possibility to represent a set of linear equations in matrix form. For example, a set of two equations with two variables. For example, the set of two equations with two variables:

$$A_{11}x_1 + A_{12}x_2 = B_1$$
$$A_{21}x_1 + A_{22}x_2 = B_2$$

may be written in compact matrix form as $AX=B$ or in full matrix form as:

$$\begin{bmatrix} A_{11} & A_{12} \\ A_{21} & A_{22} \end{bmatrix} \begin{bmatrix} x_1 \\ x_2 \end{bmatrix} = \begin{bmatrix} B_1 \\ B_2 \end{bmatrix}$$

2.2.2.3 Division

Due to the above-mentioned noncommutative properties of a matrix, the division of matrices is much more complicated than their multiplication. A full explanation can be found in books on linear algebra. Here, the related operators are described in the context of their use in MATLAB.

Identity and inverse matrices are often used in dividing operators.

An **identity** matrix I is a square matrix whose diagonal elements are 1's and the others are 0's. It can be generated with the `eye` command (see Table 2.4). In the case of multiplication of identity and square matrices, the law of commutation is retained. A by I, or I by A, yields the same result: $AI=IA=A$. For example:

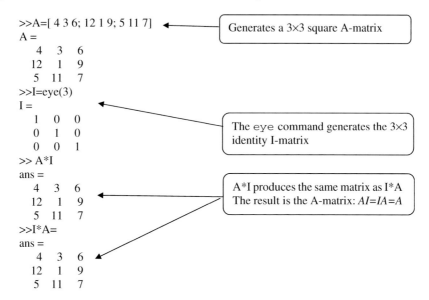

In case of multiplication of the identity matrix by a nonsquare matrix, the result is identical to multiplying this matrix by scalar 1, e.g., [4 3 6;12 1 9]*eye(3) and leads to the same product as 1*[4 3 6;12 1 9] or [4 3 6;12 1 9]*1.

Matrix B is termed **inverse** to A in the case when multiplication on the left or right leads to the identity matrix: $AB=BA=I$. The inverse to the A-matrix is typically written as A^{-1}. In MATLAB, this can be written in two ways: `B=A^-1` or with the `inv` command as `B=inv(A)`. For example, multiplying previously generated matrix A by its inverse leads to the identity matrix:

```
>> A*A^-1
ans =
    1.0000   -0.0000   -0.0000
   -0.0000    1.0000   -0.0000
   -0.0000    0.0000    1.0000
```

Where matrix manipulations are performed, the left, \, or right, /, division is often used. For example, to solve the matrix equation $AX=B$, when A is a square matrix and X and B are column vectors, left division should be used: $X=A\backslash B$. By contrast, to solve the same matrix equation but rewritten in form $XC=B$, with X and B row vectors and C as a transposed matrix of A, the right division should be used: $X=B/C$.

For example, the set of linear equations:

$$-x_1 + 2x_2 = 7$$
$$5.1x_1 + 9x_2 = 11.5$$

can be rewritten in another form, with coefficients to the right of unknowns x_1 and x_2 as:

$$-x_1 + x_2 \cdot 2 = 7$$
$$x_1 \cdot 5.1 + x_2 \cdot 9 = 11.5$$

These two possible representations correspond to the two matrix forms:

$$AX = B \text{ with } A = \begin{bmatrix} -1 & 2 \\ 5.1 & 9 \end{bmatrix}, B = \begin{bmatrix} 7 \\ 11.5 \end{bmatrix}, \text{ and } X = \begin{bmatrix} x_1 \\ x_2 \end{bmatrix}$$

or

$$XC = D \text{ with } C = A' = \begin{bmatrix} -1 & 5.1 \\ 2 & 9 \end{bmatrix}, D = B' = [7\ 11.5], \text{ and } X = [x_1\ x_2]$$

The solutions for the two discussed forms use two forms of matrix division with the commands being:

```
>> A=[-1 2;5.1 9];
>> B=[7;11.5];
>> X_left=A\B
X_left =
   -2.0833
    2.4583
```

Solution, X_left, received with the left division A\B for the equation form $AX=B$; X_left is a column vector

```
>> C=A';
>> D=B';
>> X_right=D/C
X_right =
   -2.0833    2.4583
```

Solution, X_right, received with the right division D/C for the equation form $XC=D$; X_right is a row vector. Note: in this case, C is the transposed A matrix, and D is the transposed B vector

An application example with matrix division is provided in Section 2.3.4.1.

2.2.3 Array operations

The previously described Matrix operations follow the rules of linear algebra. However, there are many calculations (in particular in material sciences) where the operations are carried out by the so-called element-by-element procedure. In these cases, to avoid confusion, we use the term *array* for them. These operations can be used in multiplication, division, and exponentiation in the same way as in addition or subtraction—element-wise. Such manipulations are performed with elements in the same positions in the arrays. In contrast to matrix operations, element-wise operations are confined to arrays of equal size. They are denoted with a point typed preceding the arithmetic operator, namely:

- .* (element-wise multiplication);
- ./ (element-wise right division);
- .\ (element-wise left division);
- .^ (element-wise exponentiation).

For example, if we have vectors $v1 = [v1_1\ v1_2\ v1_3]$ and $v2 = [v2_1\ v2_2\ v2_3]$, then element-by-element multiplication v1.*v2, division v1./v2 and exponentiation v1.^v2 yield:

$$v1.*v2 = [v1_1*v2_1\ v1_2*v2_2\ v1_3*v2_3],\ v1./v2 = [v1_1/v2_1\ v1_2/v2_2\ v1_3/v2_3]\ \text{and}\ v1.\!\hat{}\,v2 = \left[v1_1^{v2_1}\ v1_2^{v2_2}\ v1_3^{v2_3}\right]$$

The same manipulations applied for two matrices $M1 = \begin{bmatrix} M1_{11} & M1_{12} & M1_{12} \\ M1_{21} & M1_{22} & M1_{23} \end{bmatrix}$ and $M2 = \begin{bmatrix} M2_{11} & M2_{12} & M2_{12} \\ M2_{21} & M2_{22} & M2_{23} \end{bmatrix}$ lead to:

$$M1.*M2 = \begin{bmatrix} M1_{11}M2_{11} & M1_{12}M2_{12} & M1_{13}M2_{13} \\ M1_{21}M2_{21} & M1_{22}M2_{22} & M1_{23}M2_{23} \end{bmatrix}$$

$$M1./M2 = \begin{bmatrix} M1_{11}/M2_{11} & M1_{12}/M2_{12} & M1_{13}/M2_{13} \\ M1_{21}/M2_{21} & M1_{22}/M2_{22} & M1_{23}/M2_{23} \end{bmatrix}$$

And

$$M1.\!\hat{}\,M2 = \begin{bmatrix} M1_{11}^{M2_{11}} & M1_{12}^{M2_{12}} & M1_{13}^{M2_{13}} \\ M1_{21}^{M2_{21}} & M1_{22}^{M2_{22}} & M1_{23}^{M2_{23}} \end{bmatrix}$$

Element-wise operators are frequently used for calculating a function given as a series of values of its arguments. Examples of array operations are:

```
>> A1=[2 -5;11.3 4;5.2 7.1]
A =
    2.0000   -5.0000
   11.3000    4.0000
    5.2000    7.1000
```
Generates 3×2 array `A1`

```
>> A2=[-1 2.7;3 10.2;0.9 0.44]
A2 =
   -1.0000    2.7000
    3.0000   10.2000
    0.9000    0.4400
```
Generates 3×2 array `A2`

```
>> A1.*A2
ans =
   -2.0000  -13.5000
   33.9000   40.8000
    4.6800    3.1240
```
Element-by-element multiplication of `A1` by `A2`

```
>> A1./A2
ans =
   -2.0000   -1.8519
    3.7667    0.3922
    5.7778   16.1364
```
Element-by-element division of `A1` by `A2`

```
>> A2.^2
A./B
ans =
   -2.0000   -1.8519
    3.7667    0.3922
    5.7778   16.1364
```
Element-by-element exponentiation of `A2`. As a result, each term in `A2` is a power of 2

```
>> A1*A2
??? Error using ==> mtimes
Inner matrix dimensions must agree.
```
Matrix multiplication. Error as `A1` and `A2` have different inner dimensions since the row number in `A1` is not equal to the column number in `A2`

```
>> x=1:4
x =
     1     2     3     4
```
Generates four-element vector `x`

```
>> y=4+exp(-.5*x.^2)

y =
    4.6065    4.1353    4.0111    4.0003
```
Calculates vector $y=\exp(-0.5*x_2)$ at `x` previously given as four-element vector using element-by-element operations

```
>> y=(x+5)./(3*x.^2-1)
y =
    3.0000    0.6364    0.3077    0.1915
```
Calculates vector $y=(x+5)/(3x_2-1)$ at `x` previously given as four-element vector using element-by-element operations

2.2.4 Commands for generation of some special matrices

Matrices with some certain or random values can be generated with special commands. The

`ones(m,n)` and `zeros(m,n)`

commands are used for matrices of *m* rows and *n* columns with 1 and 0 as all elements.

Many engineering problems, including those related to descriptive statistics, material physics, testing, measurements technique, material informatics, properties of materials, and various simulations, involve random numbers for which the following generators of pseudorandom numbers should be used:

`rand(m,n)` or `randn(m,n)`

The former command yields a uniform distribution of elements between 0 and 1 with the latter a normal one with mean 0 and standard deviation 1. For generating a square matrix (n x n), these commands can be abbreviated to `rand(n)` and `randn(n)`. Examples are:

```
>> ones(2,4)
ans =
    1   1   1   1
    1   1   1   1
```
Generates a 2×4 matrix in which all elements are equal to 1

```
>> zeros(4,2)
ans =
    0   0
    0   0
    0   0
    0   0
```
Generates a 4×2 matrix in which all elements are equal to 0

```
>> a=rand(2,4)
a =
    0.8147   0.1270   0.6324   0.2785
    0.9058   0.9134   0.0975   0.5469
```
Generates a 2×4 matrix a with uniformly distributed random numbers between 0 and 1

```
>> v=rand(1,4)
v =
    0.8147   0.9058   0.1270   0.9134
```
Generates a row vector v with four uniformly distributed random numbers between 0 and 1

```
>> b=randn(2,4)
b =
    0.5377  -2.2588   0.3188  -0.4336
    1.8339   0.8622  -1.3077   0.3426
```
Generates a 2×4 matrix b with normally distributed random numbers

```
>> w=randn(4,1)
w =
    0.5377
    1.8339
   -2.2588
    0.8622
```
Generates a 4×1 column vector w with three normally distributed random numbers

To generate the integer random numbers, the `randi`-command can be used, as shown in Table 2.4.

Note: when we repeatedly use the `rand`, `randn`, or `randi` command, the new random numbers are generated each time. To restore the settings of the random number generator to produce the same random numbers as if you had **restarted** MATLAB, the `rng default` command should be typed and entered into the Command Window. For example:

```
>> rng default
>> a=rand(2,4)
a =
    0.8147    0.1270    0.6324    0.2785
    0.9058    0.9134    0.0975    0.5469
>> a=rand(2,4)
a =
    0.9575    0.1576    0.9572    0.8003
    0.9649    0.9706    0.4854    0.1419

>> rng default

>> a=rand(2,4)
a =
    0.8147    0.1270    0.6324    0.2785
    0.9058    0.9134    0.0975    0.5469
```

- The `rng default` command puts the starting generator settings to their default values
- The `rand` command generates the 2×4 matrix `a` of random numbers
- New `rand`, without a `rng` previously inputted, produces 2×4 matrix `a` with the new random number values
- The new `rng default` command restores the generator settings
- Restores matrix `a` generated with the `rand` command and previously inputted `rng default` command

In addition to the commands described in the previous sections, MATLAB has many others that can be used for manipulation, generation, and analysis of matrices and arrays, some of which are listed in Table 2.4.

The matrices discussed so far, except for some of those in Table 2.4, have numerical elements even if they are written as expressions as these expressions yield numbers when evaluated. However, a string/s can be used as a single element or number of elements of a matrix. A string is an array of characters—letters, digits, and other symbols. A string is entered in MATLAB between single quotes, e.g.:

Table 2.4 Commands for matrix manipulations, generation, and analysis (alphabetically).

MATLAB command	Purpose	Example (inputs and outputs)
`cross(a, b)`	Calculates cross product of 3D-vectors $\mathbf{a} = a_1\hat{i} + a_2\hat{j} + a_3\hat{k}$, $\mathbf{b} = b_1\hat{i} + b\hat{j} + b\hat{k}$: $c = \mathbf{a} \times \mathbf{b} = (a_2b_3 - a_3b_2)\hat{i} + (a_3b_1 - a_1b_3)\hat{j} + (a_1b_2 - a_2b_1)\hat{k}$ Vectors **a** and **b** must be three elements each	`>>a=[3 1 5];` `>>b=[4 2 7];` `>>c=cross(a,b)` `c =` `-3 -1 2`
`det(A)`	Calculates the determinant of square matrix A	`>>A=[4 -2 6;1 8 2;6 11.1 7];` `>>det(A)` `ans =` `-96.2000`
`diag(x)`	Generates a matrix with elements of vector x placed diagonally	`>> x=1:2:5;diag(x)` `ans =` `1 0 0` `0 3 0` `0 0 5`
`dot(a,b)`	Calculates the scalar product of two vectors with *n* elements each: $a \cdot b = a_1b_1 + a_2b_2 + \ldots + a_nb_n$; the result is a scalar	`>>a=[3 1 5];` `>>b=[4 2 7];` `>>c=dot(a,b)` `c =` `49`
`length(x)`	Returns the number elements (length) of vector x	`>> x=[2.9 5.7 -1 3];` `>>length(x)` `ans =` `4`
`max(a)`	Returns a row vector with maximal numbers of each column in the matrix a. If a is a vector, it returns the maximal number in a	`>> a=[5 8; 1 2; 7 3];` `>> b=max(a)` `b =` `7 8` `>> a=[5 8 1 2 7 3];` `>> b=max(a)` `b =` `8`
`mean(a)`	Returns a row vector with arithmetical mean values calculated for each column of the matrix a. If a is a vector, it returns the average value of the vector a	`>> a=[5 8; 1 2; 7 3];` `>> b=mean(a)` `b =` `4.3333 4.3333` `>> a=[5 8 1 2 7 3];` `>> b=mean(a)` `b=` `1`

Continued

Table 2.4 Commands for matrix manipulations, generation, and analysis (alphabetically)—cont'd

MATLAB command	Purpose	Example (inputs and outputs)

`median(a)`	Returns a row vector with median values (central values of ordered data in a) for each column of the matrix a. If a is a vector, it returns the average value of the vector a	`>> a=[5 8; 1 2; 7 3];`
		`>>b=median(a)`
		`b =`
		`   5   3`
		`>> a=[5 8 1 2 7 3];`
		`>>b=median(a)`
		`b=`
		`   4`

`min(a)`	Analogous to max but for minimal element	`>> a=[5 8; 1 2; 7 3];`
		`>> b=min(a)`
		`b =`
		`   1   2`
		`>> a=[5 8 1 2 7 3];`
		`>> b=min(a)`
		`b =`
		`   1`

`num2str(a)`	Converts a single number or numerical matrix elements into a string representation. Frequently used to combine strings and numbers within the same vector/matrix	`>> a=21.64356;`
		`>> v=[ 'a='  num2str(a)]`
		`v =`
		`'a=21.6436'`

`randi(imax,m,n)`	Returns an m by n matrix of integer random numbers from value 1 up to imax—maximal integer value	`>> randi(11,2,4)`
		`ans =`
		`   3   9  10   6`
		`   7   3   9  11`

`repmat(a,m,n)`	Generates the large matrix containing m × n copies of a	`>> A=[2 1;1 2];`
		`>> b =repmat(A,1,2)`
		`b =`
		`   2   1   2   1`
		`   1   2   1   2`

`reshape(a,m,n)`	Returns an m-by-n matrix whose elements are taken column-wise from a. Matrix a must have m*n elements	`>>a=[5 8; 1 2; 7 3]`
		`a =`
		`   5   8`
		`   1   2`
		`   7   3`
		`>>reshape(a,1,6)`
		`ans =`
		`   5   1   7   8   2   3`

Table 2.4 Commands for matrix manipulations, generation, and analysis (alphabetically)—cont'd

MATLAB command	Purpose	Example (inputs and outputs)

`size(a)`	Returns a two-element row vector; the first element is the number of rows in matrix a, while the second is the number of columns	`>>a=[5 8; 1 2; 7 3];`
		`>>size(a)`
		`ans =`
		`3 2`

`[y,ind] = sort(a, dim,mode)`	For vector or matrix. Sorts elements of a in the `dim` dimension: 1- along columns and 2- along rows and in "ascend" (default) or "descend" mode order. It returns ordered y and indices of the a for each y	`>> a=[5 8; 1 2; 7 3];`
		`>> [y,ind]=sort(a,2,'descend')`
		`y =`
		`8 5`
		`2 1`
		`7 3`
		`ind =`
		`2 1`
		`2 1`
		`1 2`

`std(a)`	Calculates standard deviation. $\left[\frac{1}{n-1}\sum_{i=1}^{n}(a_i-\mu)^2\right]^{\frac{1}{2}}$, with μ as a mean of n elements for each column of a. If a is a vector, it returns the standard deviation value of vector	`>> a=[5 8; 1 2; 7 3];`
		`>> std(a)`
		`ans =`
		`3.0551 3.2146`
		`>> a=[5 8 1 2 7 3];`
		`>> std(a)`
		`ans =`
		`2.8048`

`strvcat (t1,t2, t3,...)`	Generates the matrix containing the text strings t1, t2, t3, ... as rows	`>> t1 = 'Metal';`
		`>> t2 = 'Ceramic';`
		`>> t3 = 'Composite';`
		`>> strvcat(t1,t2,t3)`
		`ans =`
		`3×9 char array`
		`'Metal '`
		`'Ceramics '`
		`'Composite'`

`sum(a)`	Returns a row vector with column sums of matrix a. If a is a vector, it returns the average value of vector a	`>> a=[5 8; 1 2; 7 3];`
		`>> sum(a)`
		`ans =`
		`13 13`

Based on Burstein, L. (2015). *Matlab in Quality Assurance Sciences*. Cambridge: Elsevier-WP.

```
>> chr='Materials science is understanding of nature, properties and use of materials'
chr =
    'Materials science is understanding of nature, properties and use of materials'
>> str='Hardness'
str =
    'Hardness'
```
The 'Materials science ...' sentence is assigned to the chr variable

Assigns the string 'Hardness' to the variable str; it is composed of 8 letters and is an 8-element row vector (single quotes are not counted)

```
>> length(str)
ans =
    8
```
The length of the str vector is 8, not 10, as the single quotes are not counted in the string length

In computing, each character of the string is treated and stored as a number. Thus the set of characters represents a vector or an array and can be addressed as an element of a vector or array, e.g., str(5) in the string Hardness is the letter 'n'. Single quotes are not counted in the string length. Some examples with string manipulations are:

```
>> str(6)
ans =
    'e'
```
The sixth element of the vector str is the letter e; thus str(6) is e

```
>> str(5:7)
ans =
    'nes'
```
The fifth, sixth and seventh elements of the vector str are the letters n, e, and s

```
>> str([3 6 4])
ans =
    'red'
```
The 3rd, 6th and 4th elements of the vector str are the letters r, e, and d

To display string without quotes, the disp command can be used, e.g.:

```
>>disp(chr)
Materials science is understanding of the nature, properties, and use of materials
>> disp(str([3 6 4]))
red
```

Individual characters and/or whole strings can be placed as elements in a vector or matrix. Rows of the string vector/matrix are divided by a semicolon (;) just as numerical rows; strings within the rows are divided by a space or comma. String rows should have the same number of elements. Each column element must be the same length as the longest of the rows. To achieve the character alignment between string rows, the spaces should be added to shorter strings; for example:

>> Name=['Temperature';'Pressure';'Conductivity']
Dimensions of matrices being concatenated are not consistent.

Error message appeared due to inequality in length of the words: the number of the characters in the word `Temperature` *and in* `Pressure` *is 11 and 8 while in* `Conductivity` *– 12*

>> Name=['Temperature ';'Pressure ';'Conductivity'] ;disp(Name)
Temperature
Pressure
Conductivity

The one space was added after `Temperature` *and four after* `Pressure`*; now all the word strings are the same length, therefore they are successfully form the* `Name` *matrix and displayed as a column vector*

Calculating the number of spaces to add to each string in the column is a tedious procedure for the user. To avoid this, use the `strvcat` command as shown in Table 2.4.

2.2.5 Displaying table with the `disp` and `fprintf` commands

As stated in Section 2.1.7, the `disp` and `fprintf` commands can display vectors, matrices, and a caption. Thus, these commands can be used for output data in tabular form.

Let's show first how it can be done with the `disp` command. For example, air compressibility factor data with captions as per Table 2.3 should be displayed using the following commands:

42 A MATLAB® Primer for Technical Programming in Materials Science and Engineering

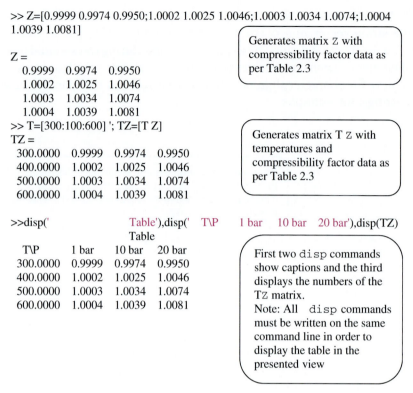

Use now the `fprintf` command. This command permits a formatted output, for example, the T numbers (first column of the TZ-matrix) can be displayed without decimal digits while the others—with four decimal digits:

```
>> fprintf('              Table\n T\P  1 bar   10 bar  20 bar\n'),fprintf('%5.0f %8.4f %8.4f %8.4f\n', TZ')
```

	Table		
T\P	1 bar	10 bar	20 bar
300	0.9999	0.9974	0.9950
400	1.0002	1.0025	1.0046
500	1.0003	1.0034	1.0074
600	1.0004	1.0039	1.0081

The first `fprintf` command shows two caption lines and the second displays the numbers of the TZ matrix.
Note: all `fprintf` commands must be written on the same command line in order to display the table in the view presented

The `fprintf` command prints rows as columns. For this, the TZ matrix was transposed for the correct table view. To display each matrix row on the new line, the \n signs (back slash and letter n without space) must be written at the end of the column format.

2.2.6 Application examples

2.2.6.1 Volume via vectors

The volume of a face-centered cubic lattice cell can be defined with the dot and cross commands as:

$$V = \left| \vec{a_1} \cdot (\vec{a_2} \times \vec{a_3}) \right|$$

where $\vec{a_1}, \vec{a_2},$ and $\vec{a_3}$ are tree-element vectors. Each *a* is the lattice constant, and the |•| and × signs designate the dot and cross products (see Table 2.4), respectively.

Problem: Calculate *V* when $\vec{a_1}=(0.5\mathbf{j}+0.5\mathbf{k})a$, $\vec{a_2}=(0.5\mathbf{i}+0.5\mathbf{k})a$, $\vec{a_3}=(0.5\mathbf{j}+0.5\mathbf{k})a$ while the *a* value is 3.567 Å (diamond). Print the defined vectors and volume with the `fprintf` commands.

The following steps should be taken to solve the problem:

- Enter *a* and generate three three-element vectors named a1, a2, and a3;
- Calculate V with the `dot` and `cross` commands;
- Display results with the two `fprintf` command. For this:
  - Create the matrix named a123 that includes the a1, a2, and a3 vectors as the matrix rows;
  - Create matrix A that combines the numbers 1, 2, and 3 (For further displaying *a* with trailing number) and matrix A with defined a1, a2, and a3 vector elements;
  - Define the a123 values as integer and a1, a2, a3 values as a fixed point with three decimal digits;
  - Transpose matrix A with the first `fprintf` command since it displays rows as columns;
  - Display text and defined *V* value, e.g., so: Volume = 11.3462 Angstrom.

The commands are:

```
>>a=3.567;
>>a1=[0 0.5 0.5]*a;
>>a2=[0.5 0 0.5]*a;
>>a3=[0.5 0.5 0]*a;
>>a123=[a1;a2;a3];
>>A=[(1:3)' a123];
```
Generates the three-element vectors a1, a2 and a3 for given value a

Generates vector A containing numbers 1, 2, and 3 before the a1, a2, and a3 vectors, which are used for the `fprintf` command

```
>>v=dot(a1,cross(a2,a3));
```
Calculates v by the given expression via the `dot` and `cross` commands

```
>>fprintf('a%1d=[%6.3f %6.3f %6.3f]\n',A')
>>fprintf('Volume=%8.4f Angstrom\n',v)
a1=[ 0.000   1.784   1.784]
a2=[ 1.784   0.000   1.784]
a3=[ 1.784   1.784   0.000]
Volume=11.3462 Angstrom
```
The first `fprintf` command displays three lines with the a1=, a2= and a3= characters and values of the defined elements of these vectors; the second `fprintf` displays the word `Volume=`, volume value, and the word `Angstrom`

2.2.6.2 Generating table of names of statistical measures

Data analysis has become an important part of materials science with the following statistical measures being calculated: mean, median, mode, range, and standard deviation.

Problem: Generate and display the matrix in which the first column is the serial number while the second is the name of the statistical measure.

To execute, use the following steps:

- Generate a numerical column of values from 1 to 5;
- Generate a string column with the names Mean, Median, Mode, Range, and Standard deviation;
- Combine these two columns into the same matrix. In MATLAB, all matrix elements must be of the same type, e.g., if strings are written in one column, then another column must also contain strings or vise-versa; in our case, we use the `num2str` command (Table 2.4), which transforms numerical data into strings:
- Display the table entitled "Statistical measures";
- Display a two-column matrix with serial numbers and measure names.

The commands are:

```
>> No=[1:5]';
```
Generates the `No` column vector of the serial numbers

```
>> Name=strvcat(' Mean',' Median',' Mode',' Range',' Standard deviation');
```
Generates the `Name` matrix with column of strings; before each word. insert a space for the separation of the serial number and the word in the next command

```
>> Table=[num2str(No) Name];
```
Generates the `Table` string matrix with columns of serial numbers and strings of the statistical measure names. The `num2str` command used to transform numbers into strings

```
>> disp(' '),disp(' Statistical measures'),disp(Table)
   Statistical measures
   1  Mean
   2  Median
   3  Mode
   4  Range
   5  Standard deviation
```
The first `disp` displays blank line; the second `disp` displays the title; and the third `disp` displays the statistical measure names

2.2.6.3 Diffusion coefficient statistics

In a laboratory, the diffusion coefficient $D \cdot 10^6 \text{ cm}^2/\text{s}$ was measured for two solute-solvent pairs, ten times for each pair. The results are presented in two rows (different pairs) and ten columns as follows:

```
3.44 3.40 3.47 3.38 3.39 3.40 3.31 3.49 3.44 3.40
4.20 4.20 4.15 4.02 4.15 4.21 4.13 4.17 4.15 4.07
```

Problem: Find the mean (arithmetic), median (central value of ordered set), range (difference between maximal and minimal values), and standard deviation for every row of data and display the mean, median, and range values as integer numbers and the resulting standard deviation with two decimal places using the `fprintf` command.

The steps are as follows:

- The diffusion coefficient data are assigned to a two-column matrix;
- The mean, median, range, and standard deviation are calculated for every column with the appropriate MATLAB commands (see Table 2.4);
- The obtained statistics are displayed using the `fprintf` command; the mean, range, and standard deviation values are shown as scientific numbers with two decimal digits.

The commands are:

```
>>D=[3.44 3.40 3.47 3.38 3.39 3.40 3.31 3.49 3.44 3.40
    4.20 4.20 4.15 4.02 4.15 4.21 4.13 4.17 4.15 4.07]*1e-6';
```
Enters two rows with the diffusion coefficient data and converts them into two columns of the D matrix

```
>>average=mean(D);
>>medi=median(D);
>>range=max(D)-min(D);
>>st_dev=std(D);
```
Calculates the mean, median, range, and standard deviation

Displays the title and obtained diffusion coefficient statistics by the `fprintf` command

```
>> fprintf('Diffusion coefficient,cm^2/s\n Mean    %5.2e %5.2e\n Median  %5.2e
%5.2e\n Range   %5.2e %5.2e\n St. dev. %5.2e %5.2e\n',...
    average,medi,range,st_dev)
```

Diffusion coefficient, cm^2/s
Mean   3.41e-06 4.15e-06
Median   3.40e-06 4.15e-06
Range   1.80e-07 1.90e-07
St. dev. 5.09e-08 6.00e-08

2.2.6.4 Mean time between failures

The technical index, which defines the average time between interruptions in a semiconductor circuit, caused by electro-migration, is the mean time to failure, MTTF. This parameter can be calculated by Black's equation:

$$MTTF = \frac{A}{j^n} e^{\frac{E_a}{kT}}$$

where E_a is the activation energy, T operating temperature, k Boltzmann constant, j current density, A is a coefficient, and n is dimensionless power.

Problem: Write the commands for obtaining the matrix *MTTF* by giving vectors *j* and *T*. Arrange output as a table whose rows represent *MTTF* at constant j and the columns at constant *T*. Use the following parameters: $E_a = 1.1$ eV, $k = 8.6173 \cdot 10^{-5}$ eV/K, $n = 2$ dimensionless, $A = 2.19 \cdot 10^{-14}$ h (cm²/A)^n, *j* is given from 1.65 to 1.8 in steps of 0.05 A/cm² and *T* is 293.15, 298.15, 300, 302, and 305 K.

The steps are as follows:

- Generate the column-vector j (size 4×1) and row-vector T (size 1×5) separately;
- Calculate the *MTTF*-matrix according to the formula. The first multiplier of the *MTTF*-expression, A/j^n, has the size 4×1 while the second, $e^{\frac{E_a}{kT}}$, of size 1×5. Thus, their product according to the linear algebra rule would be 4×5 (the extreme values of the sizes of these vectors).
- Display the title "Mean time to failure" with the `disp` command and the *MTTF*-matrix with the `fprintf` command showing the digits before the decimal point only.

The commands for the solution are:

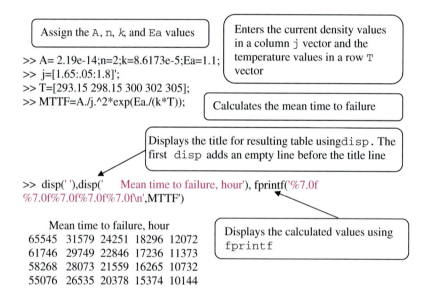

2.3 Flow control

A computational program is merely a sequence of commands that are implemented in the order in which they are written. However, there are many cases when the order of commands must be changed when they are executed as in when the calculation must be repeated with new parameters or when one expression out of several is selected to calculate a variable. For instance, the processed data were provided with two different expressions, each of which is accurate within a different area. In this case, the area should be checked to use the correct computing expression. In other situations, in calculations with the given accuracy of the answer, it may be necessary to repeat commands several times until the error in the answer diminishes to a size that is smaller than the required one. Flow control is applied to realize these processes. For these purposes, MATLAB uses special commands, usually called conditional statements. These commands direct the computer to choose which command should be executed next. The most frequent flow control commands are described below.

2.3.1 Relational and logical operators

Flow control operations are realized using relational and logical commands. Both groups of commands test the similarity between pairs of values or statements, but the former operates mostly with numerical values while the latter with Boolean values.

2.3.1.1 Relational operators

Operators comparing a pair of values are called relational or comparison operators. The application result of such an operator is written as 1 (true value) or 0 (false value), e.g., the expression $x<4$ results in 1 if x less than 4, while in 0 if otherwise.

The relational operators are:

< (less than),
> (greater than),
<= (less than or equal to),
>= (greater than or equal to),
== (equal to),
~= (not equal to).

Note: two-sign operators must be written without spaces, e.g., two equal signs must be written without space between them: ==.

When a relational operator is applied to a matrix or an array, it performs element-by-element comparisons. The comparisons return the array of 1's where the relation is true (the array has the same size as the size of compared matrix) and 0's where it is not. If one of the compared operands is scalar and the other is a matrix, the scalar is matched against every element of the matrix. The ones and zeroes are logical data types, which is not the same as numerical data type although they can be used in arithmetical operations in the same way as numerical data.

Some examples are:

```
>> 2*3==12/2
ans =
  logical
   1
```
Since 2 multiplied by 3 is identical with 12/2 the result is true. Thus, the answer is 1

```
>>cos(2*pi)~=0
ans =
  logical
   1
```
Since COS(2π)=1 and not 0, the result is true. Thus the answer is 1

```
>> M=[-6.2 8.3 -14.9;7 -8 4;-2 -15 -2]
M =
   -6.2000    8.3000  -14.9000
    7.0000   -8.0000    4.0000
   -2.0000  -15.0000   -2.0000
```
Produces a 3×3 matrix M

```
>> B=M<=0
B =
  3×3 logical array
   1  0  1
   0  1  0
   1  1  1
```
Checks whether each element in the matrix M is less than or equal to 0. Assigns results to matrix B with logical 1's in places where M<=0 and logical 0's when M>0

```
>> M(B)
ans =
   -6.2000
   -2.0000
   -8.0000
  -15.0000
  -14.9000
   -2.0000
```
Displays the elements of M, which are less than 0. Logical 1's (elements of B) are used here as addresses of elements greater than 0

```
>> M(M<0)=0
M =
        0    8.3000         0
   7.0000         0    4.0000
        0         0         0
```
Assigns zeros to the elements of M that are less than 0. Logical 1's, defined with M<0, are used here as addresses of elements less than 0

2.3.1.2 Logical operators

Logical operators are designed for operations with true or false values within the logical expressions. Similar to relational operators, they can be used as addresses in another vector, matrix, or array; see, for instance, the last two example commands.

In MATLAB, there are three logical operators: & (logical AND), | (logical OR) and ~ (logical NOT). Like the relational operators, they can be used as arithmetical operators and with scalars, matrices, and arrays. Comparison is element-by-element with logical 1 or 0 when the result is true or false; respectively, MATLAB also has equivalent logical functions: and(A,B) is equivalent to A&B, or(A,B)—to A|B, not(A,B)—to A~B. If the logical operators are performed on logical variables, the results are according to the rules of Boolean algebra. In operations with logical and/or numerical variables, the results are logical 1 or 0.

Some examples are:

```
>> x=-3.1;
```
Defines variable x

```
>> -4<x<-1
ans =
  logical
   0
```
The statement leads to an incorrect answer as it runs from left to right. -4<-3.1 is true (1), making 1<-1 is false (0)

```
>> x>-4&x<-1
ans =
  logical
   1
```
Here the logical & is used and leads to a correct result. First the inequalities are run; both are true (1); then the & is executed and leads to the answer 1

```
>> ~(x<4)
ans =
  logical
   0
```
x<4 is stated in the parentheses and executed first, is true (1), then ~1 is false, so the result is 0

```
>> ~x<4
ans =
  logical
   1
```
Here ~x is executed first, x is nonzero, then true (1), ~1 is 0, and 0<4 is true

The result of logical or relational operators depends on the order in which they are executed. The order in which combinations of relational, logical, and conditional operators is executed (so-called precedence rules) are—from the highest to the lowest—parentheses, exponentiation, NOT (~), multiplication/division, addition/subtraction, relational operators, AND (&), OR (|). The order of execution necessary for such an operator can also be reached using parentheses.

Among the MATLAB logical functions is the find command, which in its simplest forms reads as:

    i=find(x)   or   i=find(A>c)

where i is a vector of the place addresses (indices) where non-zero elements of the x (first form) are located or are elements of A larger than c (second form; in this case, any of the relational operators can also be used, e.g., <, >=, etc.); for example:

```
>> T=[11 0 5.5 0 -1.5];
```
Assigns vector T

```
>> i1=find(T)
i1 =
   1  3  5
```
Returns the addresses of the non-zero elements of the vector T; addresses are assigned to vector i1

```
>> i2=find(T<5)
i2 =
   2  4  5
```
Returns the addresses of the T vector with elements less than 5; addresses are assigned to vector i2

Application example: Screening of the metals in respect to their density
The density for ten metals appears in the following table:

Metal	Density (g/cm³)
Magnesium	1.77
Aluminum	2.71
Titanium	4.47
Chromium	6.92
Zinc	7.14
Iron/Steel	7.86
Nickel	8.89
Copper	8.94
Silver	10.49
Gold	19.02

Problem: Use the relational and logical operators to form a list of metals with densities that are: (a) less than 3; (b) between 3 and 8; and (c) greater than 8. Display the density together with metal names for each of these groups.

The steps are as follows:

- Assign the metal name and density values as they appear in the table.
- Input the names of the metals as character matrix and metal densities as a column vector, each with the names `Metal` and `Density`. Metal names have different lengths, but each row of the matrix must be the same length. To do this, each name must be supplemented with spaces to the longest name. (In this problem, this name is "iron/steel"). In addition, to create a gap between the name and density, one space character can be written at the end of each name.
- Assemble the vectors in the `Metal_Density` matrix. Note that these vectors are of different types: the `Metal` vector comprises strings while the `Density` vector comprises numbers. Thus, the latter should be transformed to the string type using the `num2str` command.
- The row indices designate the locations where the densities are less than 3 g/cm³ and are identified with the `find` logical command; assign them to the `d_less3` vector.
- Calculate the number of metals with densities of less than 3 g/cm³ by the `sum` command with the condition `density <3` (relational operator).
- Display the name of the metal and its density with the `Metal_Density` matrix where the indices are the `d_less3` vector and all column indices are assigned by the colon operator (:).
- The row indices, the number of metals with molar weights between 3 and 8 and above 8 g/cm³, and their name-density columns are defined and displayed in the same way as densities of less than 3 g/cm³.

The commands used to solve this problem are:

```
>>Metal=['Magnesium ';'Aluminum ';'Titanium ';'Chromium ';'Zink        ';
'Iron/Steel ';...
        'Nickel    ';'Coper    ';'Silver   ';'Gold     '];
>>Density=[1.77 2.71 4.47 6.92 7.14 7.86 8.89 8.94 10.49 19.02]';
>>Metal_Density=[Metal num2str(Density)];
```
Defines the matrix and vector: the Metal and Density with the metal names and density data

Creates a Metal_Density matrix that combines the metal names and density data

```
>> d_less3=find(Density<3);
```
Finds addresses (indices) where the density values are less than 3

```
>> n_d_less3=sum(Density<3)
n_d_less3 =
    2
```
Calculates the amount of metals with density values less than 3

```
>> disp(Metal_Density (d_less3, :))
Magnesium  1.77
Aluminum   2.71
```
Displays the metal names and their densities from those that are less than 3

Finds addresses where the density values are between 3 and 8

```
>> d_between3and8=find(Density>=3&Density<=8) ;
```

Calculates the amount of metals with density values between 3 and 8

```
>> n_d_between3and8=sum(Density>=3&Density<=8)
n_d_between3and8 =
    4

>> disp(Metal_Density(d_between3and8,:))

Titanium    4.47
Chromium    6.92
Zink        7.14
Iron/Steel  7.86
```
Displays the metal names their and densities from those that are between 3 and 8

Finds addresses where the density values are larger than 8

```
>> d_above8=find(Density>8);
>> n_d_above8=sum(Density>8)

n_d_above8 =
    4
```
Calculates the amount of metals with desity values greater than 8

```
>> disp(Metal_Density (d_above8, :))
Nickel    8.89
Coper     8.94
Silver   10.49
Gold     19.02
```
Displays the metal names and their densities from those that are greater than 8

2.3.2 The `If` statements

Flow control commands are used to manage the order of the execution of commands. This order is provided by various conditional statements. The first is the `if` statement, which has three forms: `if ... end`, `if ... else ... end` and `if ... else if ... else ... end`. Each `if` construction should terminate with the word `end`. Its words appear on the screen in blue. The `if` statement forms and their constructions are shown in Table 2.5:

In this table, the conditional expression uses the relational and/or logical operators, such as `a<=v1&a>=v2` or `b ~ = c`.

When the `if` conditional statement is typed and entered in the Command Window next to the prompt `>>`, the new line (and additional lines, after pressing enter) appear without the prompt until the final `end` is typed and entered.

An application example with the `if` statement is presented at the end of Section 2.3.4.2.

Table 2.5 Different forms of the If statement.

The If statement form	How the commands should be designed
`if ... end`	`if` conditional expression ... ... ⎫ ... ⎬ MATLAB® command/s ... ⎭ `end`

`if ... else ... end`	`if` conditional expression
	 ...
	 ... ⎫ MATLAB® command/s
	 ... ⎭
	 `else`
	 ...
	 ... ⎫ MATLAB® command/s
	 ... ⎭
	 `end`

`if ... else if ... else... end`	`if` conditional expression
	 ...
	 ... ⎫ MATLAB® command/s
	 ... ⎭
	 `elseif` conditional expression
	 ...
	 ... ⎫ MATLAB® command/s
	 ... ⎭
	 `else`
	 ...
	 ... ⎫ MATLAB® command/s
	 ... ⎭
	 `end`

2.3.3 Loops in MATLAB

To repeat one or a group of commands, the loop commands are used. Loop is another method of flow control. Each cycle of commands execution is termed a pass. There are two loop commands in MATLAB: `for ... end` and `while ... end`. These words appear on the screen in blue. Similar to the `if` statement, each `for` or `while` construction must terminate with the word `end`.

The loop statements are written in a general form in Table 2.6:

In `for ... end` loops, the commands written between `for` and `end` are repeated k times, a number that increases in every pass by the addition of the `step`-value. This process continues until k reaches or exceeds the final value.

The square brackets in the expression for k (Table 2.6) mean that k can be assigned as a vector, such as `k=[3.5 -1.06 1:2:6]`. The brackets can be omitted if there are only colons in k, e.g., `k =1:3:10`. Immediately after the last pass, the command following the loop is executed. In some calculations implemented the `for ... end` loops, they can be replaced by matrix operations. In such cases, the latter are actually superior as the `for...end` loops work slowly. This advantage is negligible for short loops with a small number of commands but appreciable for large loops with numerous commands.

The `while ... end` loop is used where the number of passes is not known in advance at the loop start. MATLAB executes the commands written between the `while` and `end` in each pass and the passes are repeated until the conditional expression is true. An incorrectly written loop may continue indefinitely, for example

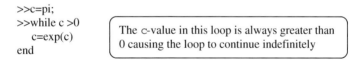

```
>>c=pi;
>>while c >0
    c=exp(c)
end
```

The c-value in this loop is always greater than 0 causing the loop to continue indefinitely

Table 2.6 Commands for loop generation.

Command form	How the commands should be designed

`for ... end`	`for k =[initial : step : final]`
	 ...
	 ... command/s MATLAB®
	 ...
	 `end`

`while ... end`	`while` conditional expression ... ... command/s MATLAB® ... `end`

54 A MATLAB® Primer for Technical Programming in Materials Science and Engineering

In this loop, after the c value exceeds the number $1.7977 \cdot 10^{308}$ (the maximum possible positive real number), the expression c = inf appears repeatedly on the screen. To interrupt the loop, the Ctrl and C keys should be pressed simultaneously.

Examples of for ... end and while ... end loops are given below for calculating exponential e^{-x} (which is present in many expressions of material science and engineering) via the Taylor series $\sum_{k=0}^{n}(-1)^k \frac{x^k}{k!}$ at $x=0.5236$:

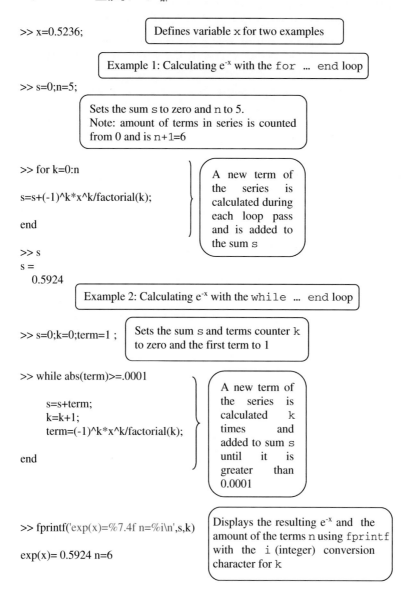

The first example is written with the `for ... end` loop. At the beginning of the first pass, the s value is equal to zero. During this pass, the first term (k=0) is calculated and added to the s. In the second pass, k=k+1=1, the second term of the series is calculated and added to the previous s value. This procedure is repeated in this example up to k=n=5. After this, the loop ends with the obtained value displayed by typing and entering of the variable names. In the case of the `for ... end` loop, the number of passes is fixed.

The second example for the `while ... end` loop presents a more complicated situation. In this case, the condition for ending the cycle must be specified. It is assumed here that the value of the kth term may not be greater than 0.0001. As in the previous case, in the first pass, the s and k values are equal to zero (k in this case is called counter) with the term value being 1 (for k=0). These values are assigned before the loop. In the first pass, the first term of the series is checked if it is greater than 0.0001. If it is true, the term added to the sum s. Then the k value is increased by 1 followed by calculation of the new term value. The value (without a sign) is checked if it is greater than 0.0001. If this condition is true, the next pass is started for the next term calculation. If it is false, the loop ends and the `fprintf` command displays the obtained s value and number of times the term was added to s. As the latter value is an integer, the conversion character i is used (within the `fprintf` command) for displaying the value of k.

Note: the `for ... end` and `while ... end` loops and `if` statements can incorporate additional loops and/or `if`-statements. The order and number of these inclusions are not restricted but are predetermined only by calculation purposes.

2.3.4 Application examples

2.3.4.1 Isothermal liquid density: Defining coefficients of the linear fit

Fitting data with some mathematical expression is a widely used technique in the material sciences to describe possible relationships between the dependent y and independent x variables. For example, an experiment shows the following data for liquid density (y) at pressures (x) at constant temperature 37.8°C (100 F):

Density, g/cm^3 = 0.8595 0.8493 0.8301 0.8166 and 0.8076
Pressure, psi = 26.003, 20.184, 12.372, 6.045 and 2.022.

This data can be described by the linear equation (termed also as linear regression) $y = a_1 + a_2 x$ in which the coefficients a_1 and a_2 can be obtained from the following set of equations:

$$a_1 n + a_2 \sum_{i=1}^{n} x_i = \sum_{i=1}^{n} y_i$$

$$a_1 \sum_{i=1}^{n} x_i + a_2 \sum_{i=1}^{n} x_i^2 = \sum_{i=1}^{n} x_i y_i$$

where n is the number of the observed values, x and y are pressure and density, respectively.

This set can be represented in the matrix forms $AX = B$ or $XA = B$, in the first case:

$$\begin{bmatrix} n & \sum_{i=1}^{n} x_i \\ \sum_{i=1}^{n} x_i & \sum_{i=1}^{n} x_i^2 \end{bmatrix} \begin{bmatrix} a_1 \\ a_2 \end{bmatrix} = \begin{bmatrix} \sum_{i=1}^{n} y_i \\ \sum_{i=1}^{n} x_i y_i \end{bmatrix}$$

and in the second case:

$$\begin{bmatrix} a_1 & a_2 \end{bmatrix} \begin{bmatrix} n & \sum_{i=1}^{n} x_i \\ \sum_{i=1}^{n} x_i & \sum_{i=1}^{n} x_i^2 \end{bmatrix} = \begin{bmatrix} \sum_{i=1}^{n} y_i & \sum_{i=1}^{n} x_i y_i \end{bmatrix}$$

Problem: Define the *a*-coefficients with left- and right-division; print the result as a linear equation with the relevant coefficients a_1 and a_2.

The following steps are to be taken:

– Generate two-row vectors with the given x (densities) and y (pressures) values;
– Generate a 2×2 matrix $A = \begin{bmatrix} n & \sum_{i=1}^{n} x_i \\ \sum_{i=1}^{n} x_i & \sum_{i=1}^{n} x_i^2 \end{bmatrix}$ —the first matrix forms; the `sum` command can be used for sums here and below;
– Generate a column vector $B = \begin{bmatrix} \sum_{i=1}^{n} y_i \\ \sum_{i=1}^{n} x_i y_i \end{bmatrix}$ —the first matrix form;
– Use left division $A \backslash B$ to calculate the *a*-coefficients;
– Display the coefficients in the written linear equation using the `fprintf` command;
– Use right-hand division A/B. Write $A = \begin{bmatrix} n & \sum_{i=1}^{n} x_i \\ \sum_{i=1}^{n} x_i & \sum_{i=1}^{n} x_i^2 \end{bmatrix}$, as per the second matrix form (the same as in the first case); and $B = \begin{bmatrix} \sum_{i=1}^{n} y_i & \sum_{i=1}^{n} x_i y_i \end{bmatrix}$ is the row vector in case of this form. The right division in this case is simply a verification of the previous solution;
– Display the a-coefficients directly in the written linear equation using the `fprintf` command.

The commands for the solution are:

```
>>x=[26.003 20.184 12.372 6.045 2.022];
>>y=[0.8595 0.8493 0.8301 0.8166 0.8076];
```
Generates vectors x and y with the density-pressure data

```
>> A=[length(x) sum(x);sum(x) sum(x.^2)];
>> B=[sum(y);sum(y.*x)];
```
The first matrix form: AX=B. Generating the matrix A and vector B

```
>> a=A\B
a =
    0.8033
    0.0022
```
Solution using left division: X=A\B

Using the `fprintf` for displaying results with 4 digits after the decimal point

```
>> fprintf('\n  The equation is y=%5.4f+%5.4f*x\n',a(1),a(2))
 The equation is y=0.8033+0.0022*x
```

```
>> aa=B'/A
aa =
    0.8033  0.0022
```
The second matrix form: XA=B. Solution using right division: X=B/A. Note, the column vector B should be transposed by the quote operator (') to the tow vector

```
>> fprintf('\n  The equation is y=%5.4f+%5.4f*x\n',aa(1),aa(2))
 The equation is y=0.8033+0.0022*x
```

2.3.4.2 Bulk modulus of a material

Measured at different applied pressures, P, the bulk modulus of a material, K, is described by the linear equation:

$$K = 550.8 + 10.1P$$

with pressures in the range 0–8 GPa:

$$0 \leq P \leq 8$$

Problem: write commands that calculate and display K values in the allowed pressure range and displays the message: "Pressure out of allowed range"; set the following P values—0, 4, 8, and 12.

The steps required are:

- Assign a 1×4 vector with the P values;
- Use the `for... end` loop in which every loop pass the new P value is defined by its index (address)—$P(i)$;
- Introduce the `if ... else ... end` statement in the loop where the current value of P is checked and the value of K is calculated when P is in the allowed range but not calculated when P is out of range. The calculated K is displayed on each loop pass by the `fprintf` command in the first case, and message "P is XXX.X GPa – out of range" is displayed in the latter.

These steps are implemented with the following commands:

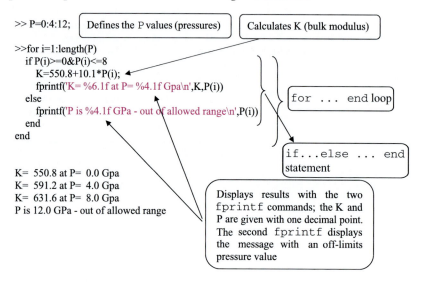

2.3.4.3 Molar concentration

A certain substance (solute) is diluted in a solution whose molar concentration (molarity), M_2, can be calculated through the following expression:

$$M_2 = \frac{M_1 V_1}{V_2}$$

where M_1 and V_1 are the initial molar concentration of the solution and its volume, respectively; V_2 is final solution volume.

A number of solutions were prepared with initial concentrations M_1, specifically 0.5, 1, 1.5, and 2 mol./L and in equal volumes $V_1 = 0.1$ L each.

Problem: Calculate the table of molar concentrations M_2 if each of the prepared solutions was diluted so that final solutions were 0.15, 0.3, 0.45, 0.6, and 0.75 L.

This problem can be resolved in two ways:

– with the `for... end` loops,
– without the loops, using the vectors only.

The following steps realize both possibilities:

– Assign the value of V_1 and generate the M_1 and V_2 vectors;
– Generate a preallocated zeros matrix with the number of rows and columns equal to the length of the M_1 and V_2, respectively. The preallocated matrix reduces operation time when the `for... end` loop is used (a minor improvement for small-sized matrices but a rather significant one for large ones);
– Calculate M_2 (with the expression above) in the two `for... end` loops: the external for M_1 and the internal—for V_2. Such a construction yields all values M_2 for each V_2;
– Display the calculated M_2 values with the `fprintf` in which the obtained values are presented with three digits after the decimal point.

- Realize now the second way—without loops:
  - for this, rewrite the expression in the form $M_1 V_1.(1/V_2)$ so that the element-wise division in brackets comes first and is followed by the $M_1 V_1$ matrix multiplication; $(1/V_2)$ produces a row vector with size $[1 \times 5]$. For the inner dimension equality, transform the row vector M_1 by the transpose (') to a column vector with size $[4 \times 1]$. The next multiplication by the scalar V_1 does not change the vector size. Finally, the product of the $[4 \times 1]*[1 \times 5]$ matrices is the 4×5 matrix with the M_2 values that must be the same as in the first case calculated.

```
>> V1=.1;
>> M1=.5:.5:2;
>> V2=.1:.2:.9;
```
Defines scalar V1 and two vectors M1 and V2 with initial substance volume and molarity as well as the desired final ssubstance volumes

First way of calculations: two `for ... end` loops for final substance concentrations

```
>> M2=zeros(length(M1),length(V2));
```
Generates a pre-allocated [4×5] zero matrix for the future calculated values of the substance concentrations M2

```
>> for k=1:length(M1)
   for j=1:length(V2)
   M2(k,j)=M1(k)*V1/V2(j);
   end
   end
```
Calculates the current final molar concentration

```
>>fprintf('\n    Concentration table\n'),fprintf('%5.3f %5.3f %5.3f %5.3f %5.3f\n', M2')
    Concentration table
0.333 0.167 0.111 0.083 0.067
0.667 0.333 0.222 0.167 0.133
1.000 0.500 0.333 0.250 0.200
1.333 0.667 0.444 0.333 0.267
```
Uses two `fprintf` for displaying the title and results with 3 digits after decimal point

Second way of calculations: the final substance concentrations by vectors, without loops

```
>> M2v=M1'*V1*(1./V2);
>> fprintf('\n    Concentration table\n'),fprintf('%5.3f %5.3f %5.3f %5.3f %5.3f\n',M2v')
    Concentration table
0.333 0.167 0.111 0.083 0.067
0.667 0.333 0.222 0.167 0.133
1.000 0.500 0.333 0.250 0.200
1.333 0.667 0.444 0.333 0.267
```
Uses two `fprintf` for displaying the title and the resulting table with 3 digits after decimal point

2.4 Questions and exercises for self-testing

1. To list the variables in the current workspace together with their byte size and other information, the following command should be used: (a) `lookfor`, (b) `whos`, (c) `who`. Choose the correct answer.

2. The predefined variable π in the longE format is displayed as: (a) 3.1416, (b) 3.1416e+000, (c) 3.141592653589793, (d) 3.141592653589793e+000. Choose the correct answer.
3. The $y = \frac{\ln 2.303}{\log 10}$ expression should be written in MATLAB as follows: (a) y=ln(2.303)/log(10), (b) y=log(2.303)/log10(10), (c) y=ln(2.303)/log10(10), (d) y=log(2.303)/log(10). Which of the forms is correct?
4. The M=[1:3; 4:3:10] command produces: (a) a row vector M with six numbers, (b) a column vector M with five numbers, (c) a rectangular matrix M with six numbers.
5. Which numbers are generated in the M matrix from the previous problem:
(a) $\begin{bmatrix} 1 & 2 & 3 \\ 4 & 3 & 10 \end{bmatrix}$, (b) $[1\ 2\ 4\ 3\ 10]$, (c) $\begin{bmatrix} 1 & 2 & 3 \\ 4 & 7 & 10 \end{bmatrix}$. Choose the correct answer.
6. The matrix $\begin{bmatrix} 0 & 0 & 0 \\ 1 & 0 & 0 \\ 0 & 1 & 0 \end{bmatrix}$ can be generated with: (a) `diag(1:3)`, (b) `eye(3,-1)`, (c) `ones(3,1)`, (d) `zeros(3,4)`. Choose the correct answer.
7. The position vector $r_A = 2i-3j+4k$ must be represented in MATLAB as: (a) rA=[2i -3j 4k], (b) ra=[2 3 4], (c) rA=[2 -3 4], (d) rA=[2 -3 4]'. Choose all the correct answers.
8. Check which of the answers below provide the correct result for [4 8;12 16]/[1 2;3 4] division:

(a) ans =   (b) ans =
    4 4   4 0
    4 4   0 4

9. Row vector F with numerical values can be transformed into a column vector with the following command: (a) F^{-1}, (b) inv(F), (c) 1/F, (d) F'. Choose the correct answer.
10. In the 4 × 4 matrix A, the element located in the third row and second column should be addressed as: (a) A(2,3), (b) A(10), A(23), (c) A(7), (d) A(3,2). Choose all correct numbers.
11. Write-in MATLAB notations the expressions for principal contact stress in y-direction (the problem for two axis-parallel steel cylinders pressed together):

$$\sigma_y = -\frac{2F}{\pi bL}\left\{\left[2 - \frac{1}{1+\left(\frac{z}{b}\right)^2}\right]\sqrt{1+\left(\frac{z}{b}\right)^2} - 2\frac{z}{b}\right\}$$

with $b = \sqrt{\frac{2F}{\pi L}\frac{\frac{(1-v_1^2)}{E_1}+\frac{(1-v_1^2)}{E_2}}{\frac{1}{d_1}+\frac{1}{d_2}}}$

Calculate value σ_y in psi for the following values of the expression parameters $F=100Lb$, $L=2$ and $z=0.001$ in, $v_1=v_2=0.3$, $E_1=E_2=3\cdot 10^7$, $d_1=1.5$, $d_2=2.75$.

12. From the van der Waals equation, the real gas temperature T (in K) can be calculated as:
$$T = (P + an^2/V^2)(V - nb)/(nR)$$
Write MATLAB commands to calculate T when the universal gas constant R=0.08206 L atm/(mol. K), number of moles n=2 moles, molar volume and pressures are 5 L and 9.3 atm, respectively; material constant a and b are 4.17 L² atm/mol.² and 0.037 L/mol., respectively.

13. The distance between two molecules via their coordinates can be determined with the expression:
$$d = \sqrt{(x_1 - x_2)^2 + (y_1 - y_2)^2 + (z_1 - z_2)^2}$$
Write commands to calculate d when the dimensionless coordinates of the first molecule are $x_1=0.1$, $y_1=0.02$, and $z_1=0.12$ while for the second molecule $x_2=0.02$, $y_2=0.5$, and $z_2=0.11$. Then, try to solve this problem giving coordinates molecules in two three-element coordinate vectors [x1 y1 z1] and [x2 y2 z2]. Present results with the `disp` command in two lines with the word "Distance" in the first line and with the d value in the second.

14. In various calculations of the solid material lattices, it is frequently necessary to define the length of vectors (termed magnitude) with the expression:
$$|u| = \sqrt{x^2 + y^2 + z^2}$$
The general form in which the **u** vector is represented is **u**=x**i**+y**j**+z**k**. Calculate the length $|u|$ when **u**=0.2**i**+0.7**j**-0.91**k**. Calculate the vector length using the sum command and element-by-element exponentiation.

Present result with the `disp` command in two lines with: the word "Vector length" in the first line, and with the $|u|$ value in the second.

15. The unit vector u_n of the same direction lattice vector **u**= x**i**+y**j**+z**k** is calculated by the expression:
$$u_n = \frac{x\mathbf{i} + y\mathbf{j} + z\mathbf{k}}{\sqrt{x^2 + y^2 + z^2}}$$
Calculate the unit vector u_n when **u**=7**i**-4**j**-11**k**. Display results with the `fprintf` command in the form "Unit vector = X.XXXi+X.XXXj+X.XXXk" (with i, j, and k not bold); the X signs designate stages for the x, y, and z values.

16. The silver heat capacity at constant pressure is given in the table.

T (K)	20	40	60	80	100	120
Cp (J/(mol. K))	1.647	8.419	14.27	17.87	20.10	21.54

(a) Generate the `Temp` row vector with the temperature data;
(b) Generate the `Cp` row vector with the heat capacity data;
(c) Join two vectors into a two-row matrix titled `Temp_Cp` with its first row presenting vector `Temp` and the second—vector `Cp`;
(d) Generate the `Head` two row matrix with names "T, K" and "Cp, J/(mol. K)" as the matrix first and second rows, respectively;
(e) Join the `Head` matrix and the `Temp_Cp` matrices to produce the problem table; use the `num2str` command for string representation of the numerical values.
(f) Display the table with the `disp` command.

17. One of theoretical expressions for low-temperature solid heat capacity at constant pressure is:

$$c_p = \frac{3R\left(\frac{\theta}{T}\right)^2 e^{-\frac{\theta}{T}}}{\left[1 - e^{-\frac{\theta}{T}}\right]^2}$$

– Write MATLB commands to calculate c_p (in J/(mol. K)) at temperatures $T = 10, 20, 40, 60, 80, 100$, and $120\,K$ when $\theta = 128\,K$ and universal gas constant $R = 8.314\,J/(mol.\,K)$.
– Display the results as a two-column matrix with temperatures in the first row and heat capacities in the second, using for this the `fprintf` commands that:
  (a) display in the first-row heading "T, K" and in the second—"Cp, J/(mol. K)";
  (b) present temperatures with integer digits only and heat capacities with three digits after the decimal point.

18. Write commands to calculate the dimensionless steady-state temperatures of a rectangular `axb` plate by the expression:

$$T = \frac{4}{\pi} \sum_{i=1,3,5,\ldots}^{n} \frac{\sinh(i\pi a\eta) \sin(i\pi a\xi)}{i \sinh(n\pi a)}$$

with $\eta = 1$, $a = b = 1$, $n = 11$, and $\xi = 0, 0.1, \ldots, 1$. Use the `for ... end` loop for ξ and the `sum` command for summarizing elements of the series (Σ).

Display a two-column table with ξ and T—columns and with heading "Coordinate Temperature." Use the `disp` commands for heading and the `fprinf` command for the numeric values.

19. Radioactive decay of a substance (or and isotope) is defined with the expression:

$$N = N_0 e^{-\lambda t}$$

where t is the time, λ is the decay rate constant, and N_0 and N are the initial and residual amounts of the substance.

(a) Calculate N for t=0, 15, 30, and 45 days when $\lambda = 0.25$ day^{-1}, $N_0 = 200$ g.
(b) Display the results as a two-column table: the first column t and the second N.

20. Stiffness of a solid material is characterized by the Young's modulus, E. For some materials, the E- data are presented in this table:

Material	Young's modulus (GPa)
Polyethylene	0.48
Teflon	0.5
Polypropylene	1.75
Magnesium	45
Aluminum	69
Steel	200
Silicon Carbide	450
Titanium	21.54
Copper	117

(a) Generate the Material column vector with material names; note: each name must be the same length as the longest table name and use the "space" sign for extension of the short names.
(b) Generate the Young column vector with Young's modulus data; sort the data in ascending order using the sort command with output that includes the sorted vector and sorted row indices.
(c) Join two vectors into the Material_Young two-column matrix when its first column presents vector Material in sorted order (use for that the indices defined in step (b)) and the second—the sorted Young vector;
(d) Display the table with the disp commands so that the table appears with sorted columns and its heading as in the problem table.

21. Using the table of the problem 20, find and list the materials with Young's modulus (a) less than 40 GPa, (b) between 40 and 100 GPa and (c) more than 100. Display the results so that the title appears for each group and then appears in two data columns (material name and Young's modulus) with the title for each column; add empty line between group of data with the disp command, including the one space sign only—disp(' '); the resulting material name-density data should be sorted in the order of ascending density.

22. Measured at 1 atm, the viscosity of the liquid propane is: temperatures $T = 240, 300, 340, 400, 440$, and 500 K, dynamic viscosity $\mu = 7.110, 8.196, 9.261, 10.82, 11.83$, and 13.29 µPa s. The data can be fitted by the Reynolds type equation $\mu = \mu_0 e^{-a_2 T}$ represented as $\ln \mu = a_1 - a_2 T$ (where $a_1 = \ln \mu_0$). The fitting coefficients a are obtainable from the following set of the linear equations:

$$a_1 n + a_2 \sum_{i=1}^{n} T_i = \sum_{i=1}^{n} \ln \mu_i$$

$$a_1 \sum_{i=1}^{n} T_i + a_2 \sum_{i=1}^{n} T_i^2 = \sum_{i=1}^{n} T_i \ln \mu_i$$

where n is the number of the measured points.

(a) Solve the set of equations with the left division rule A\B; for which must be generated a 2×2 matrix A with known terms of the equations left side (n, $\sum_{i=1}^{n} T_i$, $\sum_{i=1}^{n} T_i^2$) and column vector B with the right-side elements of the equation set;

(b) Solve the set of equations with the right division rule B/A, for which must be generated the same matrix A as previous step and row vector B with the right-side elements of the equation set;

(c) Display the derived coefficients with 3 digits after the decimal point and the resulted $\mu_{fit} = \mu_0 e^{-a_2 T}$ equation with $\mu_0 = e^{a_1}$.

23. Use equations for dynamic viscosity μ_{fit} and viscosity data μ for the problem 21 to:

(a) Calculate the absolute values of the relative percentage error $\varepsilon = 100 \left| \dfrac{\mu - \mu_{fit}}{\mu} \right|_i$ for each T_i-point;

(b) Compute the mean relative error $\bar{\varepsilon}$ and error range in values of viscosities calculated with the determined viscosity equation;

(c) Display table with T, μ, μ_{fit} and ε data; present T-values as integer and all other values with 2 digits after the decimal point.

(d) Display the following values: error $\bar{\varepsilon}$, maximal error, and error range; each with 3 digits after the decimal point.

2.5 Answers to selected questions and exercises

2.
    (c) 3.1416e+000

3.
    (c) y=ln(2.303)/log10(10)

6.
    (b) randn

8.
    (d) E'

11.
```
sigma_y =
-4.8438e+03
```

14.
```
Vector length
1.1576
```

16.

T,K	20	40	60	80	100	120
Cp,J/(mole K)	1.647	8.419	14.27	17.87	20.1	21.54

19.
```
    t, day  N, g
         0  200.0000
   15.0000  4.7035
   30.0000  0.1106
   45.0000  0.0026
```
21.
```
Materials with E less 40 GPa
Polyethylene   0.48
Teflon         0.5
Polypropylene  1.75
Titanium       21.54
Materials with E between 40 and 100 GPa
Magnesium    45
Aluminum     69
Materials with E greater than 100 GPa
Copper         117
Steel          200
Silicon Carbide 450
```
23.
```
Temperature Measured mu  Fitted mu  Error
    240       7.11        7.16      0.44
    300       8.20        8.30      0.52
    340       9.26        9.16      0.57
    400      10.82       10.61      0.63
    440      11.83       11.71      0.66
    500      13.29       13.57      0.70
Average error 0.59 Maximal error 0.70 Error range 0
```

3

Graphics and presentations

The presentation of the results of observations, tests, or calculations in the form of graphs or diagrams is widespread in engineering and sciences in general and in materials science and engineering in particular. MATLAB® has a myriad of commands for these targets. The available commands generate two- (sometimes called XY or 2D) and threedimensional (XYZ or 3D) plots.

Two-dimensional graphics allow you to create a variety of linear, nonlinear and semi- or logarithmic graphs, bars, histograms, pies, and scatter plots, to name a few options. Several curves can be plotted in one plot while several plots can be presented in a separate figure window. Constructed plots can be formatted with the desired line style or marker shape, thickness or color, with the additions of a line, grid, text, caption, or legend.

To present data with more than two variables, the plots having three axes can be used. MATLAB® provides a variety of means for visualizing three-dimensional data. These means permit the building of spatial lines, mesh, and surface plots as well as various geometrical figures and images. Generated 2D or 3D plots can be formatted with the commands or interactively from the Figure Window.

The most important commands for two- and threedimensional plotting will be presented below. The description below is based on the assumption that the reader has thoroughly studied the previous chapter of the book. Therefore, the explanations to the commands are written in most cases not in special frames (as in Chapter 2), but as inline MATLAB® comments next to the percent (%) sign.

3.1 Two- and three-dimensional plots

Graphs with two and three coordinate axes are called two- and three-dimensional plots, respectively. There are many types of these plots and, accordingly, many commands for generating these plots. Below, we describe commands generating 2D and 3D, and some specialized plots.

3.1.1 Two-dimensional plots

An important basic command used for XY plotting is the `plot` command, which in its simplest form can be written as:

`plot(y)` or `plot(x,y)`

where `x` and `y` are two vectors of equal length, the first being used for horizontal, *x*, while the second is for the vertical, *y*, axes.

The first `plot` command draws the y-vector values versus their places (indices) in the vector. After inputting the second `plot` command with the given values of x and y, the curve `y(x)` is created in the MATLAB® Figure window. In both cases, by default, a linear scale graph is generated with a blue solid line between the unmarked points.

For example, the low-temperature enthalpy changes ΔH_t^o of silver at standard pressure was measured. The obtained data is: $T=20$, 30, 40, 50, 60, 70, 80, 90, and 100 K, $\Delta H_T^o =$ 8.001, 39.14, 105.2, 206.1, 336.2, 489.5, 660.7, 815.9, and 1042 J/mol. To present these data in plot with the x-axis as temperature and the y-axis as enthalpy, we must type the following commands in the Command Window:

```
>> T=20:10:100;
>> H=[8.001 39.14 105.2 206.1 336.2 489.5 660.7 815.9 1042];
>> plot(T, H)
```

After entering these commands, the Figure window is opened with the ΔH_T^o- temperature plot as shown in Fig. 3.1.

To change the default setting for the line style and/or marker type, thickness, and color, these optional arguments may be added to the `plot` command just after the x and y identifiers:

```
plot(x,y,'Line Specifiers','Name','Value')
```

where the Line Specifiers parameter determines the line type, marker symbol, and color of the plotted lines—see Table 3.1; the pair `'Name'`–`'Value'` represents the one of the possible curve properties: the `Name` is a string with name of the desired property; and the `Value` is a number or a quoted string that specifies the property itself.

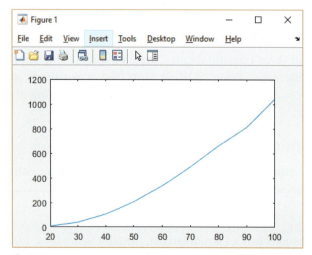

FIG. 3.1 The enthalpy ΔH_T^o of silver generated by the plot command; default plot settings.

Table 3.1 The 'Line Specifiers' arguments for the plot command.

Marker type	Specifier (marker symbol)	Plot representation with generating command/s	Line or color style	Specifier (line or color symbol)	Plot representation with generating command/s
Asterisk	*	`>> plot(1,'*')`	Dash-dot	-.	`>> plot([264,255,225],'-.')` `>> ylabel('HV')`
Circle	o	`>> plot(1,'o')`	Dashed	-- (two minuses without space between them)	`>> plot([264,255,225], '--')` `>> ylabel('HV')`
Cross	x	`>> plot(1,'x')`	Dotted	:	`>> plot([264,255,225], ':')` `>> ylabel('HV')`
Diamond	d	`>> plot(1,'d')`	Solid (default)	-	`>> plot([264,255,225], '-')` `>> ylabel('HV')`

Continued

Table 3.1 The 'Line Specifiers' arguments for the plot command—cont'd

Marker type	Specifier (marker symbol)	Plot representation with generating command/s	Line or color style	Specifier (line or color symbol)	Plot representation with generating command/s
Five-pointed asterisk	p or pentagram	>> plot(1,'p')	Black	k or black	>> plot([264,255, 225],'-k') >> ylabel('HV')
Point	.	>> plot(1,'.')	Blue (default for single line)	b or blue	>> plot([264,255, 225], '-b') >> ylabel('HV')
Plus	+	>> plot(1,'+')	Cyan	c or cyan	>> plot([264,255, 225], '-c') >> ylabel('HV')
Square	s	>> plot(1,'s')	Green	g or green	>> plot([264,255, 225], '-g') >> ylabel('HV')

Table 3.1 The 'Line Specifiers' arguments for the plot command—cont'd

Marker type	Specifier (marker symbol)	Plot representation with generating command/s	Line or color style	Specifier (line or color symbol)	Plot representation with generating command/s
Six-pointed asterisk	h or hexagram	>> plot(1,'h')	Red	r or red	>> plot([264,255, 225], '-r') >> ylabel('HV')
Triangle (inverted)	v	>> plot(1,'v')	Magenta	m or magenta	>> plot([264,255, 225],'-m') >> ylabel('HV')
Triangle (upright)	^	>> plot(1,'^')	Yellow	y or yellow	>> plot([264,255, 225], '-y') >> ylabel('HV')

Some frequently used property names with required values can be obtained in Table 3.2.

The property names and values (if string) as well as the line specifiers are typed in the `plot` commands in inverted commas. The specifiers and property names with their values can be written in any order, with the option of omitting any of them. Any omitted properties will be defined by default.

Table 3.2 Frequently used property names of the plot command, their values, and purposes.

Property name (spelling)	Purpose	Property value	Example
`LineWidth` or `linewidth`	The width of the curve line	A number in points (1 point is 1/72 inch or 0.35 mm). The default line width is half a point	`>>plot([264,255, 225],'-b',...` `'linewidth',10)` `>> ylabel('HV')`
`MarkerSize` or `markersize`	The size of the marker (by the symbol)	A number in points Default value is 6. For '.' marker—1/3 of specified size	`>> plot(1,'o','markersize',20)` `>> ylabel('HV')`
`MarkerEdgeColor` or `markeredgecolor`	The color of the marker or the edge line color for filled markers	A character in accordance with color specifiers in Table 3.1	`>>` `plot(1,'o','markeredgecolor','r',` `...` `'markersize',20)` `>> ylabel('HV')`

Table 3.2 Frequently used property names of the plot command, their values, and purposes—cont'd

Property name (spelling)	Purpose	Property value	Example
`MarkerFaceColor` or `markerfacecolor`	The fill color for markers that have closed area (e.g., circle or square)	A character in accordance with color specifiers in Table 3.1	>>plot(1,'o','markerfacecolor','g',... 'markeredgecolor','r','markersize',20) >> ylabel('HV')

Below are additional examples of the `plot` commands with various `Line Specifiers` and `Name-Value` properties:

plot(y,'-c') : generates the cyan solid line between the unmarked points, with x-equidistant point coordinate (element addresses of vector y) and y- as vertical point coordinates.
plot(x,y,'o') : generates the points with *x,y*-coordinates marked by the circle.
plot(x,y,'y') : generates the yellow solid (default) line that connects the points.
plot(x,y,'--p') : generates the blue (default) dashed line, points marked with the five-pointed asterisks.
plot(x,y,'k:h') : generates the black dotted line, points marked with the six-pointed asterisks.
plot(x,y,'-gx' ,'markersize',10,'markeredgecolor','g') : generates 0.5 point (default, 1 point has size 1/72 in. or ~0.35 mm) green solid line with 10 points marked with the green crosses (x).

plot(T,H, '-s','LineWidth',4,'MarkerSize',15,'MarkerEdgeColor','k', 'MarkerFaceColor','y') : plots as magenta 4-points solid line with T, H-values marked with the 15 points black-edged yellow squares.

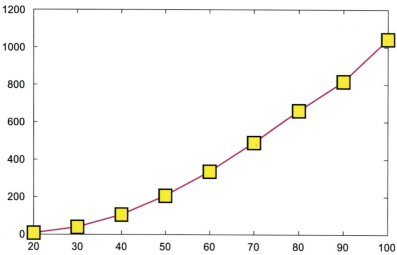

FIG. 3.2 The enthalpy-temperature data generated by the plot command with specifiers and property settings.

Entering the last command with previously used enthalpy-temperature data, we can obtain the plot shown in Fig. 3.2.

In the examples above, the plots are generated by given *x,y*-points. In this case, we can say that data is given in tabular form. In many cases, the function can be given as *y(x)* expression. In this case, the vector of *y*-values should be calculated at the given vector of *x*-values (see example in Section 3.1.2.3).

Note:

- Type and enter the `close` command in the Command Window to close one Figure window; to close more than one Figure window, use the `close all` command.
- With each input of the `plot` command, the previous plot is deleted.

3.1.1.1 Multiple curves on the 2D plot

To generate more than one curve on the same plot, at least two options can be used: a `plot` command with more than one pair of x,y-vectors or use the `hold on` command to save the generated curve and then add a new one.

The `plot` command for multiple curves

The commands for generating two or more curves in the same plot have the following forms:

`plot(x1,y1,x2,y2)`   or   `plot(x1,y1,x2,y2,...,xn,yn)`

where `x1` and `y1`, `x2` and `y2`, ... , `xn` and `yn` are pairs of equal-length vectors containing the *x,y*-coordinates of the plotting points. These commands create plots with two or n curves respectively. For example, let's generate in a single plot two curves of the specific entropy-temperature data for superheated propane vapor given at two pressures. The temperatures T are 80, 90, 100, 110, 120, 130, and 140 K. Entropies $s = $ 2.056, 2.114, 2.172, 2.228, 2.284, 2.340, 2.395 kJ/(kg K) all at pressure in 1 MPa, and $s = $ 1.867, 1.933, 1.997, 2.059, 2.119, 2.178, 2.236 kJ/(kg K) at pressure 2 MPa.

To generate a graph with two curves, enter the following commands (without comments) in the Command Window:

>> x=80:10:140;   %creates vector with temperatures, for x axis
>> y1=[2.056 2.114 2.172 2.228 2.284 2.340 2.395]; %creates vector with the 1st entropy series for y axis
>> y2=[1.867 1.933 1.997 2.059 2.119 2.178 2.236]; %creates vector with the 2st entropy series for y axis
>> plot(x,y1,x,y2,'--k')%generates two lines: solid and dashed, the latter - in black

The resulting plot with two curves is shown in Fig.3.3.

To generate more than two curves in the same plot, the new x- and y-vectors should be introduced in the `plot` command for each additional curve.

Note: If the y is given as a matrix (not a vector) with n columns, the plot(x,y) command draws n curves in the same plot in accordance with the columns of the y-matrix; e.g., the two curves shown in Fig. 3.3 can be generated by entering the >>y=[y1;y2]′ command (which joins two vectors into a two-column matrix), and then the >>plot(x,y) command.

The `hold` command for generating multiple curves

Another option for generating two or more curves in the same plot is to add a new curve to an existing plot. For this, type the `hold on` command after the first plot creation and then

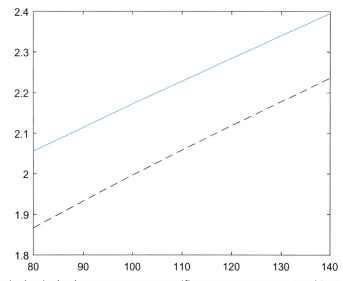

FIG. 3.3 Two curves in the single plot; propane vapor specific entropy at pressures 1 and 2 MPa.

enter a new `plot` command with the new curve coordinates. To complete the `hold on` process, enter the `hold off` command, which stops the hold process and allows the next graph to show without previously plotted curves. For example, the new series of specific entropy content: 1.689, 1.782, 1.860, 1.932, 1.994, 2.063, 2.126 in kJ/(kg K), given at 3 MPa, can be added to the existing graph (Fig. 3.3) by entering the additional commands:

```
>> y3=[1.689 1.782 1.860 1.932 1.994 2.063 2.126];   % create new vector of
                                                       entropies for y axis
>> hold on
>> plot(x,y3,':r')                                   % add dotted red line to the existing plot
>> hold off
```

Generated plot is shown in Fig. 3.4.

3.1.1.2 Several plots on the same page

It is often necessary to place multiple plots on the same page or, in other words, in the same Figure window. For this, the `subplot` command should be used; the command divides the Figure window (and page, when this window is printed) into m by n rectangular panes. It can be written in two equal forms:

`subplot(m,n,p)`   and   `subplot mnp`

where m and n are the rows and columns of the panes, and p is the plot number that this command makes current (see panes and plot numbering, Fig. 3.5A). The p within the first command form can be written as vector with two or more adjacent panes, meaning that

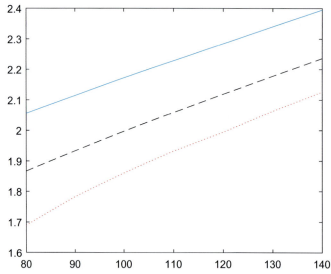

FIG. 3.4 Three curves in the same plot; the third curve (specific entropy at 3 MPa) produced with the `hold on/off` commands.

Chapter 3 • Graphics and presentations   77

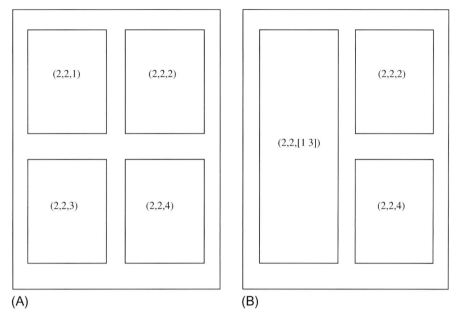

FIG. 3.5 The page arranged in four (A) and three (B) rectangular panes.

the plots can be arranged asymmetrically so that one of the plots can be placed into two or more columns or rows (see Fig. 3.5B).

For example:

subplot(2,2,3) or subplot 223: Divides the page into 4 rectangular panes arranged in 2 rows and 2 columns and makes pane 3 current.
subplot(2,2,[1,3]): Divides the page into 3 rectangular panes with second and fourth panes of usual sizes and enlarged third pane that combines the first and third panes (top and bottom left panes); the first pane is current.
subplot(2,3,3) or subplot 233: Divides the page into 6 rectangular panes arranged in 2 rows and 3 columns and makes pane 3 current.
subplot(2,1,1) or subplot 211: Divides the page into 2 rectangular panes arranged in a single column and makes the first pane current.
subplot(1,2,1) or subplot 122: Divides the page into 2 rectangular panes arranged in a single row and makes the second pane current.

As an exercise, generate three plots on one page, placed in three panes accordingly to the Fig. 3.5B:

– in both left panes, generate an ellipse by the coordinates in the parametric form $x=3\sin(t)$ and $y=5\cos(t)$ with $t = 0, \pi/50,\ldots, 2\pi$.
– in the top right pane, plot the *T,H* data (as per Fig. 3.3); make 2 points line width, 5 points marker size, black edge, and yellow face marker colors.

– in the bottom right pane, plot the *T,s* three-curve (as per Fig. 3.4); make the second and the third curves black dashed and red dash-dotted respectively.

The MATLAB® command for generating the plots:

```
>>subplot(2,2,[1,3])                              % makes panes 1 and 3 current
>>t=0:pi/100:2*pi;                                % creates the t vector
>>plot(5*sin(t),4*cos(t));                        % plots the ellipse
>>subplot(2,2,2)                                  % makes pane 2 current
>>T=20:10:100;                                    % creates the T vector
>>H=[8.001 39.14 105.2 206.1 336.2 489.5 660.7 815.9 1042];   % H vector
>>plot(T,H,'-mo','LineWidth',2,'MarkerSize', 5,'markeredgecolor', 'k',
'MarkerFaceColor','y')                            % creates the enthalpy plot
>>subplot(2,2,4)                                  % makes pane 4 current
>>T=80:10:140;                                    % creates the T vector
>>s1=[2.056 2.114 2.172 2.228 2.284 2.340 2.395];  % creates the s1 vector
>>s2=[1.867 1.933 1.997 2.059 2.119 2.178 2.236];  % creates the s2 vector
>>s3=[1.689 1.782 1.860 1.932 1.994 2.063 2.126];  % creates the s3 vector
>>plot(T,s1,T,s2,'--k',T,s3,':r')                 % generates the entropy plot
```

The resulting plot is shown in Fig. 3.6.

3.1.1.3 Formatting 2D plots using commands or the Plot Tools editor

Generally, the figure has a title, grid, axis labels, suitable axes ranges, text, legend, different colors, and various types of curve lines. These can be set by either using special commands or interactively using the Plot editor (see Section 3.1.1.4), accessible from the Figure window.

Commands for 2D plot formatting

Plot formatting commands are effective when the computer program is created (see Chapter 4); these commands should be entered after the graph has already been built with the `plot` command. The most common commands are described below.

The `grid on/off` commands

To display the grid lines on the current plot, the `grid on` command is applied. The `grid off` command hides the grid lines of the latticed plot. For example, typing `grid on` in the Command Window immediately after the Fig. 3.3 generation by the `plot`, will draw the grid lines in the figure.

The `axis` commands

These commands are used for the appearance, hiding, and scaling of the axes. Some of its possible forms are:

Chapter 3 • Graphics and presentations   79

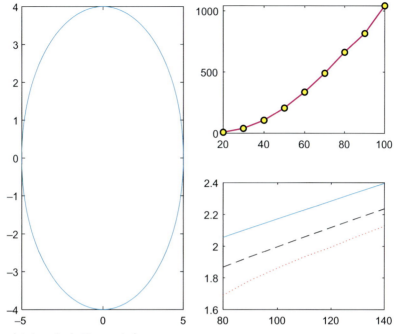

FIG. 3.6 Three plots in a single Figure window.

```
axis([x_min x_max y_min y_max])

axis equal

axis square

axis tight

axis off
```

The first form adjusts the *x* and *y* axes to the coordinate limits written as four-element vector in the brackets while the second sets the same scale for the *x* and *y* axes (the ratio *x/y*, width-to-height, is called the aspect ratio); the third sets the plot shape to be square; the fourth sets the axes limits to the range of the data to be applied to the plot; and the last removes the axis and background from the plot.

As an example, enter the sequence commands that draw the entropy plot (as in Fig. 3.3) without and with the `axis square` command. For clarity, these two plots are generated in the same Figure window with the two `subplot` commands in such a way that the plot with two entropy lines is displayed in the left pane without any of `axis` commands, and the same plot is displayed with the `axis square` command in the right pane.

80 A MATLAB® Primer for Technical Programming in Materials Science and Engineering

```
>>x=80:10:140;                          % creates vector x
y1=[2.056 2.114 2.172 2.228 2.284 2.340 2.395];    %creates vector y1
y2=[1.867 1.933 1.997 2.059 2.119 2.178 2.236];    %creates vector y2
subplot 121              % divides window in 2 pains; 1st pane is current
plot(x,y1,x,y2,'--k')                   % creates first plot
subplot 122              % divides window in 2 pains; 2nd pane is current
plot(x,y1,x,y2,'--k')                   % creates second plot
axis square                   % sets the second plot to be square
```

The results are shown in Fig. 3.7A and B.

As can be seen, when the `axis square` command is inputted (Fig. 3.7B); the axes are equal in size, and the plot becomes square.

The `xlabel`, `ylabel`, and `title` commands

These commands provide the plot with the text for the *x* and *y* axes and that which appears at the top of the plot. The desired text should be written as string, between single quotes. The commands have the forms:

$$\texttt{xlabel('text string')}$$

$$\texttt{ylabel('text string')}$$

$$\texttt{title('text string')}$$

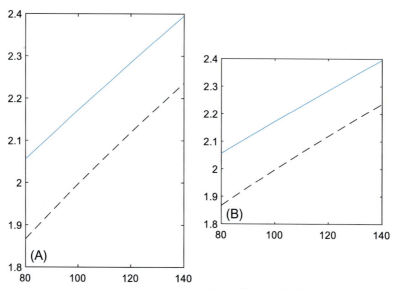

FIG. 3.7 The two entropy curves in the plots constructed without (A) and with (B) the `axis square` command.

The text string can include Latin and Greek letters. The font size of the text, its name, color, style, tilt, and some other property options can be written within each command separated by a comma from each other and from the quoted text string (see Section "Formatting text strings").

The gtext and text commands
These commands place text labels in the plot and have the following forms:

$$\text{gtext('text string')}$$

$$\text{text(x,y,'text string')}$$

The gtext command allows the user to put text interactively at the place that he chooses with the mouse. After the command is entered, the Figure window appears with two crossed lines; using the mouse, the user moves the crosshair to the proper point and then inputs the text by clicking the left mouse button.

The text command produces text starting from the point of coordinates *x* and *y*.

As an example, add the title, xlabel, ylabel, text, and grid commands to the commands used to construct the plot in Fig. 3.2:

```
>> plot(T,H,'-ms','LineWidth',4,'MarkerSize',15,'MarkerEdgeColor','k',
'MarkerFaceColor','y')
>> title(' Enthalpy vs Temperature')                    % the caption
>> xlabel('Temperature, ^oK')                           % ^o - upper o
>> ylabel(' Enthalpy, J/mole ')
>> text(40,475,' Enthalpy line ') % the 'Enthalpy line' words at x=40 and y=475
>> grid on
```

The resulting plot is shown in Fig. 3.8.

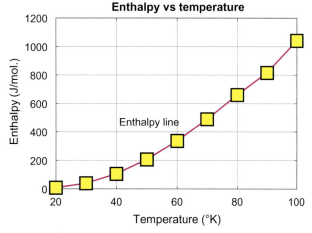

FIG. 3.8 The isobaric enthalpy plot formatted with xlabel, ylabel, title, text, and grid on commands.

82 A MATLAB® Primer for Technical Programming in Materials Science and Engineering

The `legend` command
This command is applied to print explanations for each of the plotted curves; it has the following form:

`legend('text string1','text string2',...,'Location', location_area)`

The `'text string1','text string2',...` are the explanatory text strings written within the frame. The `'Location'` property is optional; it specifies the area, `location_area`, where the explanations should be placed, for example:

`legend('line 1','line 2','Location', 'NorthEastOutside')` displays the legend frames with two line strings (`line 1`, `line 2`) placed outside the plot frames on the upper-right corner;

`legend('line 1','line 2','Location','Best')` displays the legend frames with two line strings (`line 1`, `line 2`) placed inside the plot at the best possible location (having least conflict with curve/s within the plot).

When the `legend` command is used without the `location` property, the default legend location is in the upper-right corner of the plot.

For example, input the following command to add a legend to Fig. 3.4:

>> legend('P=1 MPa','P=2 MPa','P=3 MPa','Location','Best');

After this, the plot with the legend looks like Fig. 3.9:

Formatting text strings The text string in the above commands can be formatted by writing special characters (called modifiers) inside the string immediately after the backslash, in the form `\ModifierName{ModifierValue}` (when entered, appeared in blue). Some

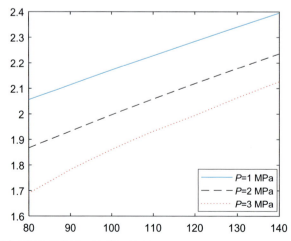

FIG. 3.9 Plot with legend for three isobaric specific entropy curves.

Table 3.3 Modifiers for the text string/s formatting.

Modifier	Purpose	Example
`fontsize{number}`	Specifies the size of the letters	`\fontsize{12}`—sets size 12 to the letters
`\fontname{name}`	Sspecifies the name of the font in use	`\fontname{Courier New}`—sets the Courier New font
\name of the Greek letter	Specifies a Greek letter	`\sigma` sets the lowcase σ `\Sigma` sets the capital Σ
\b	Specifies the bold font	`\b substance`—sets the bold font to the word/s following the modifier: **substance**
\it	Specifies the italic (tilted) font	`\it substance`—sets the word/s following the modifier to be tilted: *substance*
\rm	Specifies the normal (Roman) font	`\rm substance`—sets the Roman font to the word/s following the modifier: substance
_ (underscore)	Specifies subscripts	A_ d—the _ sign sets the d-letter to be subscripted: A_d To extend this sign to more than one letter, the curled brackets are used, e.g., 12_{dec} is displayed as 12_{dec}
^ (caret)	Specifies superscripts	^oC—the ^ symbol sets "o" to be superscripted: $°C$ To apply this sign to more than one letters, the curled brackets are used, e.g., e^{-(x-5)} is displayed as $e^{-(x-5)}$

useful modifiers for setting the font name, style, size, color, Greek letters, or sub- and superscripts are given in Table 3.3.

Another possibility for text formatting is to include the pair—property name and value—in the command just after the text string. Similar to the `plot` command (see Section 3.1), several property pairs can be included in the formatting command, and each of these pairs should be written after comma and with dividing commas in the form `'Name', Value`.

For example, the command

>> text(40,475,' Enthalpy line ', 'fontsize',12)

written for generating text in Fig. 3.8, introduces the text with size of 12 instead of 10 (default, used in Fig. 3.8).

The same effect can be achieved by writing the modifier name and value inside the string (as per Table 3.3):

>> text(40,475, ' \fontsize{12} Enthalpy line')

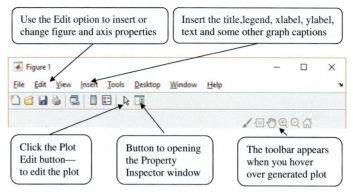

FIG. 3.10 The menu and bar buttons for plot editing from the Figure window.

3.1.1.4 About interactive plot formatting

There is an assortment of buttons and menu items in the Figure window that can be used to format a plot interactively; for this, first the Edit Plot button should be clicked. The Figure window bar line, containing the Edit Plot and other most frequently used buttons, are shown in Fig. 3.10. Properties of axes and lines, as well as the entire figure, can be changed using the pop-up menu, summoned by clicking the Edit option in the Figure menu. The title, axis labels, texts, and legend can be activated using the pop-up menu, which appears after clicking on the Insert option of the menu.

The text, legend, and other objects can be added/changed after activating "Plot Edit" mode, by clicking on the appropriate item in the Insert menu option. Property Inspector is opened using the Open Property Inspector button or by double-clicking on the shifted curve, plotted point or axes; it provides means for changing or editing various characteristics of the clicked object. For more detailed information about available plot tools, input the doc plottools command in the Command Window.

3.1.2 Three-dimensional plots

There are three main groups of commands for presenting meshes, surfaces, and lines in a three-dimensional space. These commands together with various formatting commands are described below.

3.1.2.1 Presenting line in 3D plots

In a three-dimensional space, a line is constructed from points characterized by three coordinates each, and straight lines connecting the adjacent points. Like the two-dimensional plot command, plot3 command is used for the 3D plot generation. The simplest shape of the command is:

```
plot3 (x,y,z) ,
```

a more complicated form is:

```
plot3(x,y,z,'Line Specifiers','PropertyName',PropertyValue)
```

In these commands—*x*, *y*, and *z* are equal-sized vectors containing coordinates of each of the points; the `Line Specifiers`, `PropertyName`, and `PropertyValue` have the same meaning as in the respective two-dimensional `plot` command.

The `grid on/off`, `xlabel` and `ylabel` commands, and analogously, the `zlabel` commands, can be also used to add grid and axis labels to the 3D plot.

For an example, construct a three-dimensional plot using the following commands:

```
>> t=-6*pi:pi/100:6*pi;
>> x=t;
>> y=t.*sin(2*t);
>> z=t.*cos(2*t);
>> plot3(x,y,z,'k','LineWidth',2)
>> grid on
>> xlabel('x'),ylabel('y'),zlabel('z')
```

These commands compute the coordinates by the expressions $x=t$, $y=t \cdot \sin(2t)$ and $z=t \cdot \cos(2t)$, with parameter *t* changed from -6π up to 6π with the step of $\pi/100$. The `plot3` command is used here, with the line color specifier `'k'` (black) and the line width property pair—`'LineWidth'`, 2—that increases the line width to two points (1/36 in. or ~0.7 mm); two last commands generate the grid and the captions to the axes.

The resulting three dimensional plot appears in the Figure window with the generated line in the shape shown in Fig. 3.11.

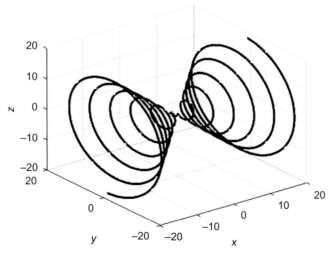

FIG. 3.11 Spiral line, narrowing and expanding, generated in three-dimensional coordinates.

3.1.2.2 Presenting mesh in 3D plots

Two basic commands generate surfaces in MATLAB®: `mesh` and `surf`. In order to understand the principles of surface reproducing, it is useful to clarify the graphical mesh construction. In general, the mesh is comprised of the points (mesh nodes) and the lines between them. Every point in 3D space has three coordinates (x, y, and z), which are used to reconstruct a surface. In actual cases, z is a function of the two variables x and y; thus it is necessary for every (x,y)-pair to obtain a z-value. The x,y coordinates of the pairs are ordered in a rectangular grid of the x,y plane. The node coordinates of this grid can be represented as two equidimensional matrices containing separately the x and y coordinates; for each grid node the z-coordinate should be calculated by the (x,y)- pair values and assembled then into a z-matrix.

An example of points represented in three-dimensional space is shown in Fig. 3.12.

The area of the x and y coordinates for which the z-coordinates must be obtained is called the domain. In the figure above, the domain represents the orthogonal grid in the x,y-plane with the limits −2 and 2 for each of the plane axes. By writing all the x-values ordered by rows (along each iso-y line), we obtain the X-matrix; as it can be seen this matrix has the same coordinates in each column. An analogous procedure yields the Y-matrix with the same coordinates in each row. The matrices are:

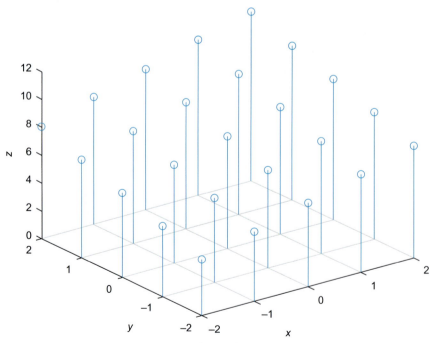

FIG. 3.12 Points (signed by "o") in three-dimensional interpretation and its x,y-plane rectangular grid. (The plot generated with the `stem3(x,y,z)` command; used expression $z = 8+x+y$.)

$$X = \begin{pmatrix} -2 & -1 & 0 & 1 & 2 \\ -2 & -1 & 0 & 1 & 2 \\ -2 & -1 & 0 & 1 & 2 \\ -2 & -1 & 0 & 1 & 2 \\ -2 & -1 & 0 & 1 & 2 \end{pmatrix} \quad Y = \begin{pmatrix} -2 & -2 & -2 & -2 & -2 \\ -1 & -1 & -1 & -1 & -1 \\ 0 & 0 & 0 & 0 & 0 \\ 1 & 1 & 1 & 1 & 1 \\ 2 & 2 & 2 & 2 & 2 \end{pmatrix}$$

The z-coordinates can be obtained for every *x,y* point of the domain using the element-by-element calculations. After the *X*, *Y*, and *Z* matrices are defined, the whole surface can be plotted.

In MATLAB®, the *X* and *Y* matrices are created by a special `meshgrid` command from the specified vectors of *x* and *y*. The command has the forms:

[X,Y] = meshgrid(x,y)   or   [X,Y] = meshgrid(x)

X and Y here are the matrices of the grid coordinates that determine the division of the domain, x and y are the vectors that represent x- and yaxis division respectively. The second command form can be used when the x- and yvectors are equal.

In case of Fig. 3.12, the obtained above *X* and *Y* matrices can be produced with the following commands:

```
>> x=-2:2;
>> [X,Y]=meshgrid(x)
X =
  -2  -1   0   1   2
  -2  -1   0   1   2
  -2  -1   0   1   2
  -2  -1   0   1   2
  -2  -1   0   1   2
Y =
  -2  -2  -2  -2  -2
  -1  -1  -1  -1  -1
   0   0   0   0   0
   1   1   1   1   1
   2   2   2   2   2
```

After the X, Y, and Z matrices are generated, the 3D mesh can be drawn. The command that generates the mesh with colored lines is:

mesh(X,Y,Z)

where the X and Y matrices are defined with the `meshgrid` command for the given vectors x and y, while the third matrix, Z, is given for every X, Y—node pair or is calculated for these matrices using the given *z(x,y)*-expression.

To summarize, to build a 3D plot, first create the grid in the *x,y* plane with the `meshgrid` command. Then calculate or define the value of *z* for each *x,y* grid node. Finally, generate the plot with the `mesh` command.

As an example, plot the 3D mesh graph for the ideal gas law equation:

$$P = nR\frac{T}{V}$$

where R is the gas constant equal to 8.3145 Pa m^3/(K mol.); n is number of moles of the gas; P is the gas pressure in Pa, T is the temperature in Kelvin degrees, V is volume in m^3. To create a graphic, define $n=1$ mole, $V=0.5 \ldots 7$ m^3 and $T=160 \ldots 300$ K, and assign ten values of V and T each.

The program that plots the 3D mesh reads:

```
>> R=8.3145;n=1;                % assigns gas constant and moles number
>> v=linspace(0.5,7,10);        % assigns v vector
>> t=linspace(160,300,10);      % assigns t vector
>> [V,T]=meshgrid(v,t);         % generates the V and T matrices
>> P=n*R*T./V;                  % calculates the P matrix
>> mesh(V,T,P)                  % generates the mesh plot
>> xlabel('Volume,m^3'),ylabel('Temperature,^oK'),zlabel('Pressure, Pa')
```

The resulting plot is shown in Fig. 3.13.

3.1.2.3 Surfaces in 3D plots

The `mesh` command, described in the previous subsection, produces a three-dimensional mesh with colored lines. Nevertheless, the surfaces between the mesh lines are white, not

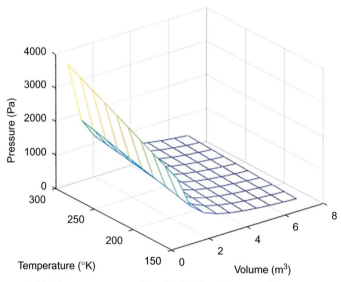

FIG. 3.13 3D plot generated by the `mesh` command for the ideal gas law equation.

colored. To generate a plot with a colored surface the, `surf` command is used, its form being:

`surf(X,Y,Z)`

X, Y, and Z being the same matrices as in the `mesh` command.

To illustrate the command implementation, we apply the commands of the previous example with the `surf` command entered in place of the `mesh`:

```
>> R=8.3145;n=1;
>> v=linspace(0.5,7,10);
>> t=linspace(160,300,10);
>> [V,T]=meshgrid(v,t);
>> P=n*R*T./V;
>> surf(V,T,P)                        % generates the surface plot
>> xlabel('Volume,m^3'),ylabel('Temperature,^oK'), zlabel('Pressure, Pa')
```

These commands generate the following plot (Fig. 3.14):
Note:

- The `surf` and `mesh` commands can be used without the X and Y matrices in the form `surf(Z)` or `mesh(Z)`. In this case, the Z values are plotted versus the row numbers (x-coordinates) and column numbers (y-coordinates) of the Z-matrix;
- The grid appears automatically in the plot when the `surf` and `mesh` commands are executed and can be removed using the `grid off` command.

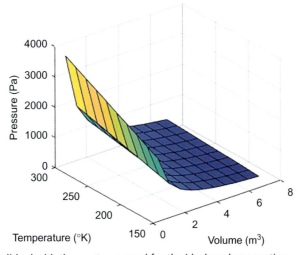

FIG. 3.14 3D plot accomplished with the `surf` command for the ideal gas law equation.

3.1.2.4 Formatting and rotating 3D plots

Many commands described for 2D plot formatting, such as `grid`, `title`, `xlabel`, `ylabel`, and `axis`, are suitable for 3D plot formatting. However, there are a variety of additional commands used for the formatting of a three-dimensional plot. Some of them are described below.

Setting the figure colors with the `colormap` command

Color plays an important role in plots, especially in three-dimensional plots; lines, meshes, and surfaces have a default color. By entering the 3D graph commands, color is automatically generated according to magnitude of the represented surface (range of z-values). Nevertheless, the colors can be set by the user with the `colormap` command:

    colormap(c)

where c is a three-element row vector with the triplet of the color values.

In the triplet, termed RGB (red-green-blue), the three numbers define the intensity of red, green, and blue colors, respectively. Intensities are graded from 0 to 1 and can take, for example, the following values:

c=[0 0 0] - Black	c=[1 0 1]	- Magenta
c=[1 0 0] - Red	c=[0.5 0.5 0.5]	- Gray
c=[0 1 0] - Green	c=[1 0.62 0.4]	- Copper
c=[0 0 1] - Blue	c=[0.49 1 0.83]	- Aquamarine
c=[0 1 1] - Cyan	c=[1 1 1]	- White
c=[1 1 0] - Yellow	c=[0.804 0.498 0.196]	- Gold

For example, the default color of the mesh lines in Fig. 3.13 can be changed to green with the `colormap([0 1 0])` command; this command should be entered after the plot creation.

Another form of the command is:

    colormap color_name

This form permits the name introduction for some common colors: the `color_name` may be `jet`, `cool`, `summer`, `winter`, `spring`, or others. For example, the `colormap summer` changes the current colors to shades of green and yellow. The default `color_name` is `parula`. You may obtain more information about available colors by entering the `doc colormap` command in the Command Window.

The `view` command, viewpoint, and different 3D projections of the graph

Previously generated 3D plots, as each 3D MATLAB® plot, are shown from a certain viewpoint. The plot orientation relative to the eye of the viewer is regulated by the `view` command with the form:

    view(azimuth,elevation)

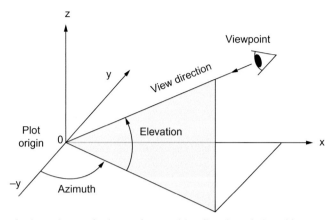

FIG. 3.15 Viewpoint and orientation angles in 3D plots; positive directions designed by *arrows*.

where azimuth and elevation are the names of the two view orientation angles shown in Fig. 3.15: the azimuth is the horizontal (*x,y*-plane) rotation angle measured relatively to the negative direction of the *y*-axes. The angle is positive in counterclockwise rotation; the elevation is the vertical angle that defines the geometric height of the observation point above the *x,y*-plane. This angle is positive when it is higher than the *x,y*-plane.

Both orientation angles must be in degrees; their default values are: azimuth = −37.5 degrees, and elevation = 30 degrees.

Various orientation angles, generated projections, and examples of the view command are presented in Table 3.4.

For an example, generate four plots of various views of the ideal gas law graph in one Figure window: the default (see previous example, Fig. 3.14), mirrored to default, the top and the front projections. The commands are:

Table 3.4 Orientation angles and corresponding projections.

Azimuth angle (degree)	Elevation angle (degree)	Projection	Example command
0	90	Top view—the *x,y*-projection (default two-dimensional view)	>>view (0,90) or >>view(2)
0	0	Side view—the x,z-projection	>>view(0,0)
90	0	Side view—the y,z-projection	>>view (90,0)
−37.5	30	Default three-dimensional view	>>view(3)
37.5	30	Mirror view of the default view	>>view (37.5,30)
180	90	Rotate the view around the x-axis by 180 degrees	>>view (180,90)

```
>>R=8.3145;n=1;
>>v=linspace(0.7,7,10);t=linspace(160,300,10);
>> [X,Y]=meshgrid(v,t);
>>P=n*R*Y./X;                  % calculates pressures by the ideal gas equation
>>subplot(2,2,1), surf(X,Y,P)           % generates the first plot, default view
>>xlabel('Volume, m^3'),ylabel('Temperature, ^oK'),zlabel('Pressure, Pa')
>>title('Default view')
>>axis tight                            % sets axis limits to the data range
>>subplot(2,2,2), surf(X,Y,P)           % generates the second plot
>>view(37.5,30)     % azimuth=37.5 and elevation=30 mirroring to the default
>> xlabel('Volume, m^3'),ylabel('Temperature,^oK'),zlabel('Pressure, Pa')
>>title('az=37.5^o, el=30^o')
>>axis tight                            % sets axis limits to the data range
>>subplot(2,2,3), surf(X,Y,P)           % generates the third plot
>>view(2)                    % azimuth =0 and elevation =90 – top view
>>xlabel('Volume, m^3'),ylabel('Temperature, ^oK')
>>title(' az =0^o, el =90^o')
>>axis tight                            % sets axis limits to the data range
>>subplot(2,2,4), surf(X,Y,P)           % generates the fourth plot
>>view(0,0)                  % azimuth =0 and elevation =0 – side view
>>xlabel('Volume, m^3'),zlabel('Pressure, Pa')
>>title(' az =0^o, el =0^o')
>>axis tight                            % sets axis limits to the data range
```

The results are shown in Fig. 3.16.

The commands above perform the following actions;

- Assign the R and n values;
- Create v and t vectors;
- Create grid matrices X and Y in the range of the x with the `meshgrid` command;
- Calculate the P pressures for each pair of the X and Y values;
- Divide the page (Figure window) into four panes and select the first pane current using the `subplot` command; generate the first plot using the `surf` command at default viewpoint;
- Set the axis limits for the data range with the `axis tight` command;
- Select the second pane current using the `subplot` command; generate the second plot with the `surf` command;
- Set viewpoint angles so that the second plot represents a mirror view (to the default view) with the `view` command;
- Set the axis limits to the data range with the `axis tight` command;
- Select the third pane for the third plot using the `subplot` command; generate the third plot using the `surf` command;
- Set the top view angles with the `view(2)` command;
- Set the axis limits to the data range using the `axis tight` command;
- Select the fourth pane for the fourth plot using the `subplot` command; generate the fourth plot with the `surf` command;

Chapter 3 • Graphics and presentations   93

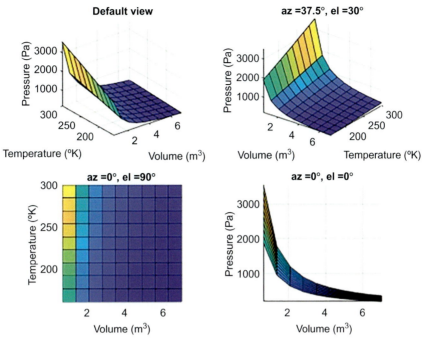

FIG. 3.16 Four different projections of the ideal gas law graph.

- Set side view angles with the view(0,0) command;
- Set axis limits to the data range using the axis tight command.

3.1.2.5 About the 3D plot rotation

A generated plot can be rotated manually. In the latest version of MATLAB®, this is done using the mouse after clicking the "rotate 3D" button, . This button, among others, appears after placing the mouse cursor on the plot in the Figure window (Fig. 3.17A). The azimuth, Az, and the elevation, El, angle values simultaneously appear in the bottom-left corner of the window. A plot view of the rotation regime is shown in Fig. 3.17B).

The plot rotation mode also can be set with the rotate3d on command. After typing and entering this command in the Command Window, execute the following:

- Go to the Figure window containing the generated plot and place the mouse cursor on it.
- By holding the left mouse button and moving the mouse, we can rotate the plot and view the azimuth Az and elevation El angle values that appear.

The rotation mode can be interrupted by entering the rotate3d off command.

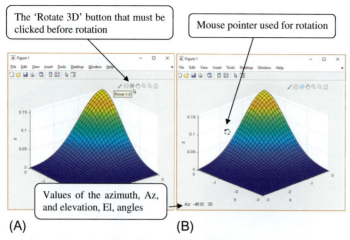

FIG. 3.17 The Figure window with a plot before (A) and during (B) rotation.

3.1.3 Specialized two- and threedimensional plots

In addition to the described 2D and 3D graphs, the material science specialists use, for example, logarithmical plots, plots with error boundaries for each x,y point, scatter plots, and some others. These graphs can be constructed using specialized commands. Some of these graphs, together with a list of additional graphic commands, are presented briefly below.

3.1.3.1 Plot with error bars

Due to method limitations and/or the equipment inaccuracy, observations of tested or investigated data is always obtained with some uncertainty. Thus, it is desirable to plot recorded values with their error limits at each measured point. Such plotting can be accomplished with the errorbar command, which represents observed points with error limits. The two simplest forms of this command are:

errorbar(x,y,l_e,u_e)   or   errorbar(x,y,e)

where *x* and *y* are the data vectors, l_e, u_e, and *e* are equal-size vectors with lower, upper and symmetrical (two-sided equal) errors, respectively.

For an example, generate the Figure window with two plots showing the elastic modulus *E* of a geopolymer measured at the different mean compressive strength σ. The data is: $E = 27.1, 29.3, 34.2, 35.0, 37.2$ GPa and σ = 48.1, 59.2, 74.3, 82.5, 89.1 MPa. For illustration, the first plot shows this data with side-asymmetric errors: the upper one is 0.05·*E*, GPa, and the lower one is 0.025·*E*, GPa. The second plot shows this data with symmetric errors of ± 0.005·*E*, GPa, at each point, yielded by entering the commands (see Fig. 3.18).

Chapter 3 • Graphics and presentations 95

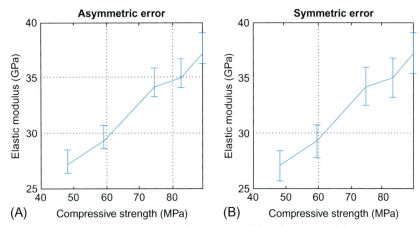

FIG. 3.18 Two plots of the elastic modulus data with asymmetric (A) and symmetric (B) errors.

```
>> E=[27.1 29.3 34.2 35.0 37.2 ];            %vector of the elastic modulus values
>> sigma=[48.1 59.2 74.3 82.5 89.1];   %vector of the compressive strength values
>> u_e= 0.05*E;                                      % vector of the upper errors
>> l_e=u_e/2;                                          % vector of the lower errors
>> e=u;                                                    % two-side error vector
>> subplot(1,2,1),                                  % makes the first pane current
>> errorbar(sigma,E,l_e,u_e)               % plot data with asymmetric errors
>> xlabel('Compressive strength, MPa'),ylabel('Elastic modulus, GPa')
>> title('Asymmetric Error'),grid on
>> subplot(1,2,2),                                  % makes the second pane current
>> errorbar(sigma,E,e)                          % plot data with symmetric error
>> xlabel('Compressive strength, MPa'),ylabel('Elastic modulus, GPa')
>> title('Symmetric Error'),grid on
```

To format the plot with error bars—the line color, style or marker—the line style, color and marker specifiers should be included in the `errorbar` command, e.g., inputting the `errorbar(sigma,E,e,'r--o')` command changes the line color to red, the line style to a dashed line, and assigns the data points with circles into the right plot of the Fig. 3.18.

3.1.3.2 Plot with semilogarithmic axes

When generating graphs in the field of MSE and technology, one of the coordinates is often necessary to be presented in a logarithmic scale. This allows the display of values in a wider range than can be shown using a linear axis; in addition, exponential relationships become linear in form in a semi-logarithmic scale. For this purpose, the `semilogy` or `semilogx` commands may be implemented. Following is the command in its simplest form:

```
semilogy(x,y,'Line Specifiers') and semilogx(x,y,'Line Specifiers')
```

The first command generates a plot with a log-scaled (base 10) y-axis and a linear x-scale; the second command generates a plot with a linear y-axis and a log-scaled x-axis; the `Line Specifiers` parameter here represents the same specifiers as used with the `plot` command that specifies the line type, plot symbol, and color for the lines drawn in the semilog plot.

For example, the data for motor oil dynamic viscosity μ is 1328, 185.31, 38.071, and 9.588 mPa s at 0 °C, 30 °C, 60 °C, and 90 °C respectively. A semilogarithmic plot can be generated with the following commands:

```
>> mu=[1328 185.31 38.071 9.588];T=0:30:90;
>> semilogy(T, mu,'-o')    % generates plot with semi-log y axis and lineal x-axis
>> xlabel('Temperature, ^oC'), ylabel('Dynamic viscosity, mPa.s')
>> grid on                                           % shows semi-log grid
```

The data on the semilog graph in Fig. 3.19 is nearly linear; on a normal graph produced by the `plot` command, this data generates an exponential-like curve.

3.1.3.3 Plot with two y-axes

Frequently, two curves that should be placed in the same graph have different y-scale and/or different units. For generating the two different vertical axes in the same plot, use the two `yyaxis` commands

`yyaxis left` and `yyaxis right`

these commands activate the left-side and right-side y-axis of the plot, respectively. After inputting each of these commands, the `plot` and desired formatting commands must be entered to for each y-axis.

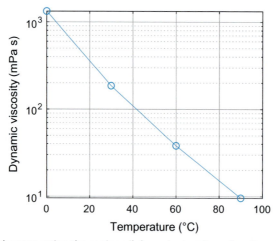

FIG. 3.19 The semilog graph representing the engine oil dynamic viscosity as function of temperature.

As an example, generate in the same plot both the dynamic viscosity and the density of an engine oil as function of the temperature. The viscosity is 1328, 582.95, 185.31, 38.071, and 9.588 mPa s, at temperatures 0°C, 10°C, 30°C, 60°C, 90°C respectively, while the density is 0.8916, 0.8725, 0.8539, and 0.8359 g/cm^3 at temperatures 0°C, 30°C, 60°C, and 90°C respectively. Place the left y-axis for viscosity and the right y-axis for density on the graph. The desired plot can be generated with the following commands:

```
>>mu=[1328 582.95 185.31 38.071 9.588];
>>T_mu=[0 10 30:30:90];                    % vector x for the left axis
>>ro=[0.8916 0.8725 0.8539 0.8359];
>>T_ro=0:30:90;                            % vector x for the right axis
>>yyaxis left                              % activates left y-axis
>>plot(T_mu,mu,'o-')                       % left y-axes plot, circled points
>>xlabel('Temperature, ^oC')
>>ylabel('Dynamic viscosity, mPa.s')
>>yyaxis right                             % activates right y-axis
>>plot(T_ro,ro,'s-')                       % right y-axes plot, squared points
>>xlabel('Temperature, ^oC')
>>ylabel('Density, g/m^3')
>>grid on
```

After entering these commands, the resulting plot with two different y-axes is (Fig. 3.20):

Note: the `xlabel` and `grid on` commands are entered only once since the x-axis and the grid are common to the two y-axes plot. This is also true for all other two-dimensional plot formatting commands not related to the y-axes.

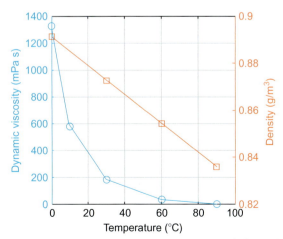

FIG. 3.20 The plot with two different y-axes representing the dynamic viscosity (left y-axis) and density (right y-axis) of engine oil.

3.2 Statistical plots

MATLAB® has numerous commands for statistical processing and visualization of the experimental or theoretical data of a broad range of the material science/engineering problems, e.g., in processing experimental data, describing the mechanical and thermodynamic properties of polymers or in visualizing material physics describing the distribution of particle velocities, etc. Most statistical means are available in the Statistical toolbox, which is beyond the scope of the book. Here we describe some commands and the Data Statistics tool that I available in the basic MATLAB®.

3.2.1 The `hist` command

A histogram, one of the most popular bar graphs, represents the data distribution and is used for statistical analysis of the data. To construct a histogram, the total data range is divided into several equal subintervals called bins, after which the quantity of values falling into each of these subintervals is counted. Defined quatities, termed frequencies, are graphically presented as heights of the bars; the subintervals are presented as bar widths. A histogram is plotted using the `hist` command. The following is the simplest form of this command:

```
hist(y)
```

where `y` is the vector containing the data points; the command generates a graph of bins that present the numbers of the data points in each of the 10 (default) equally-spaced bins.

For example, the experimental data of the molecular weight, in kDa, of 30 polymer chains are: 513, 499.5, 499.13, 480050, 525470, 496670, 482140, 533780, 494380, 485270, 492660, 478800, 472000, 563150, 541390, 507690, 468570, 478360, 495590, 519790, 466700, 441750, 463770, 508340, 509780, 511290, 496740, 504590, 488100, 504590, 488100, and 521550. To plot a histogram based on the cited data, enter the following commands:

```
>>w=[513 499 499 480 525 496 482 533 494 485 492 478,...
472 563 541 507 468 478 495 519 466 441 463 508,...
509 511 496 504 488 504 488 521];            % vector of weights, kDa
>>hist(w)                                    % plots histogram
>>xlabel('Weight, kDa')
>>ylabel('Number of weights per one bin')
```

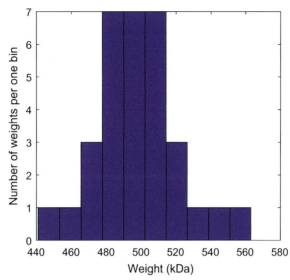

FIG. 3.21 Histogram plot of the polymer weight data.

The resulting plot is shown in Fig. 3.21.
Another available form of the hist command is:

[n,xc]=hist(y)

which outputs the n and xc vectors with the numbers of data points falling to each of the bins and *x*-centers of the bins.

Note: this command yields a numerical output but does not plot a histogram.

Using the [n,xc]=hist(y) command for the example above, the following frequencies can be displayed in the Command Window:

```
>> [n,xc]=hist(w)
n =
  1 1 3 7 7 7 3 1 1 1
xc =
 Columns 1 through 5
   447.1   459.3   471.5   483.7   495.9
 Columns 6 through 10
   508.1   520.3   532.5   544.7   556.9
```

There are additional forms of the hist command that can be applied to define bin locations or to input desired *x* values for the each bin center; for more detailed information, enter the doc hist command in the Command Window.

3.2.2 The `bar` command

The histogram can be plotted with the `bar` command when the bin centers and the frequencies are known. The command displays the values of a vector/matrix as vertical bars. Two simplest forms of the command are:

`bar(x,y) and bar(x,y,color)`

where x and y are vectors/matrices with data values, e.g., using the previous example data, x is the weights and y is the number of weights per one bin; color specifies the color of the bar and is the same as for the `plot` command (see Section 3.1.1); default triplet of the color is [0 0.45 0.74].

For an example, represent the *x*-centers of the polymer weights and weights frequencies obtained in the previous example in two plots generated in the one Figure window: the first with the `bar(x,y)` command and the second with the `bar (x,y,color)` command.

To build these plots, enter the following commands:

```
>> xc=[447.1 459.3 471.5 483.7 495.9 508.1 520.3 532.5 544.7 556.9];   %vector of bar centers
>> n=[1 1 3 7 7 7 3 1 1 1];                    % vector of frequencies
>> subplot(1,2,1)
>> bar(xc,n)                                   % sets the [0 0.45 0.74] (default) bar color
>> xlabel('Weight, kDa')
>> ylabel('Frequences, units')
>> subplot(1,2,2)
>> bar(xc,n,'g')                               % sets the green bar color
>> xlabel('Weight, kDa')
>> ylabel('Frequences, units')
```

Generated plots are shown in the Fig. 3.22.

The `bar` command has more graphical capabilities than the `hist` command in plotting statistical data. In addition, this command can be used not only in statistical representations but also to display trends or compare different groups of data or to track changes over time. Detailed information about the available command forms can be obtained by inputted the `doc bar` command in the Command Window.

3.2.3 The Data Statistics tool

The Data Statistics tool is a graphical interface that calculates and visualizes the entered data set together with some quantitative indices describing this data (such data processing is termed descriptive statistics). The representing indices are: minimum value, maximum value, mean, median, mode, standard deviation, and range (the maximum minus minimum values). The steps for using the Data Statistics tool are described below using

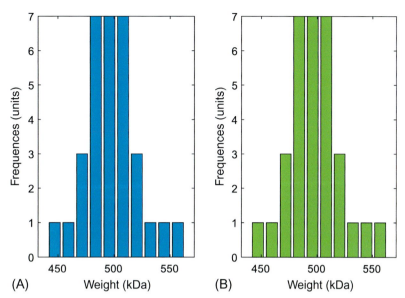

FIG. 3.22 Bar plots of the polymer weight data generated without (A) and with (B) the color specifier of the bar command

the dataset of the polymer molecular weight discussed in the previous example (Sections 3.2.1 and 3.2.2).

In the first step, the data must be entered and plotted with the following commands:

```
>>w=[513 499 499 480 525 496 482 533 494 485 492 478,...
   472 563 541 507 468 478 495 519 466 441 463 508,...
   509 511 496 504 488 504 488 521];
>>plot(w)
```

In the second step, select the Data Statistics line of the Tools menu button—Fig. 3.23A; the small-scale Data Statistics window appears with statistics for the ordinate (x) and abscissa (y) values.

The third step is to mark the boxes with the desired statistical indices; the lines showing the selected indices—Fig. 3.24 (all possible indices selected to be shown in this plot).

In the fourth step, you may transfer the defined values of the statistical indices to the MATLAB® workspace; click for this the "Save to workspace …" button (see Fig. 3.23B) and mark in the small appeared pane the desirable (x and/or y) statistics for transferring to the workspace.

Note:

the range cannot be displayed on the plot;
in this example, the x-axis represents the point numbers (indices) and are not subject to statistical description.

102 A MATLAB® Primer for Technical Programming in Materials Science and Engineering

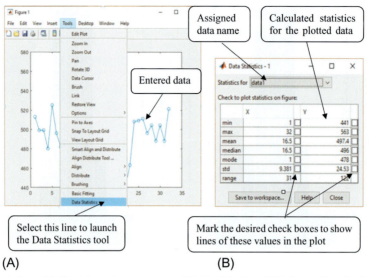

FIG. 3.23 Figure window with the data and popup menu of the Tools button (A) before calling the Data Statistics tool; the Data Statistics window (B) after calling.

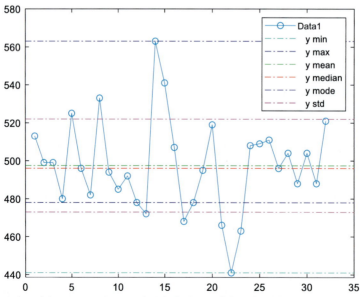

FIG. 3.24 Resulting plot of the Data Statistics tool with the lines of the selected statistical indices that describe this data.

3.3 Supplementary commands for generating 2D and 3D graphs

In addition to the described graphical commands, many other commands are available in MATLAB® for 2D and 3D plotting. A complete list can be obtained by entering the `help graph2d`, `help graph3d` or `help specgraph` commands in the Command Window. Table 3.5 presents some further commands for two- and threedimensional plotting

Table 3.5 Supplementary 2D and 3D plotting commands.

Commands and their purposes	MATLAB examples	Generated plots
`barh(x,y)` displays the values in a vector or matrix as horizontal bars	`>> x=0:10:100;` `>> y=[ 1,2,1,2,5,2,5,2,3,1];` `>> barh(x,y)` `>> xlabel('Number of Grades per Bin')` `>>ylabel('Grades')`	
`bar3(Y)` generates 3D-bar plot by the data grouped in columns	`>> Y = [2 4 8 5 6 3 10 3 6];` `>> bar3(Y)`	
`box on/off` displays axes outline/returns the axes to its default view	`>> sphere(25),axis equal` `>> box on` `>>boxoff`	

Continued

Table 3.5 Supplementary 2D and 3D plotting commands—cont'd

Commands and their purposes	MATLAB examples	Generated plots

`contour(X,Y,Z,v)` displays the x,y-plane of iso-lines of the Z(X,Y)-surface; Z is interpreted as height with respect to the x,y-plane; v is the number of contour lines. The form `c=contour(X,Y,Z,n)` with addition the `clabel(c)` command displays the level values c of the iso-lines	`>> x=linspace(-2,2,20);`
	`>> [X,Y] = meshgrid(x);`
	`>> Z=2e3*X.*Y.*exp(-X.^2-Y.^2);`
	`>> contour(X,Y,Z,6)`
		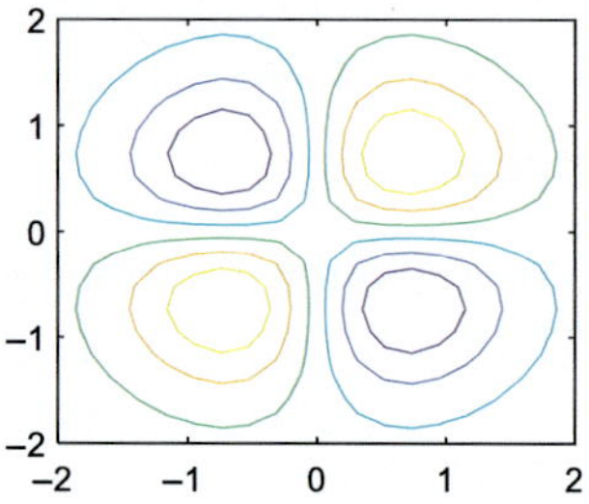

	`>>% or with level values`
	`>> c = contour(X,Y,Z,6)`
	`>> clabel(c)`
		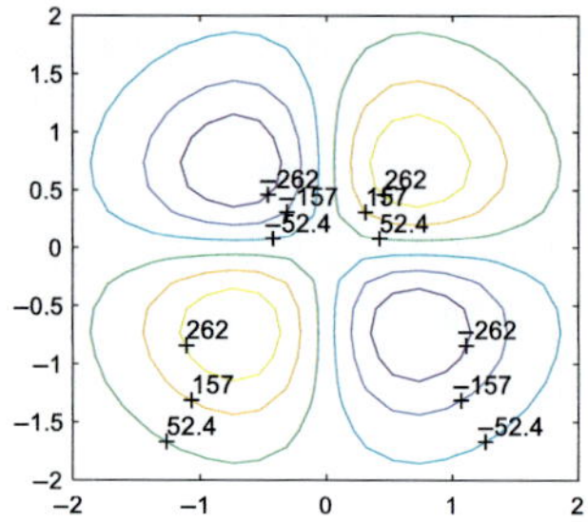

`contour3(X,Y,Z,n)` displays x,y-planes of iso-lines of the Z(X,Y) surface in 3D; n as the number of contour lines	`>> x=linspace(-2,2,20);`
	`>> [X,Y] = meshgrid(x);`
	`>> Z=X.*exp(-X.^2-Y.^2);`
	`>> contour3(X,Y,Z,6)`
		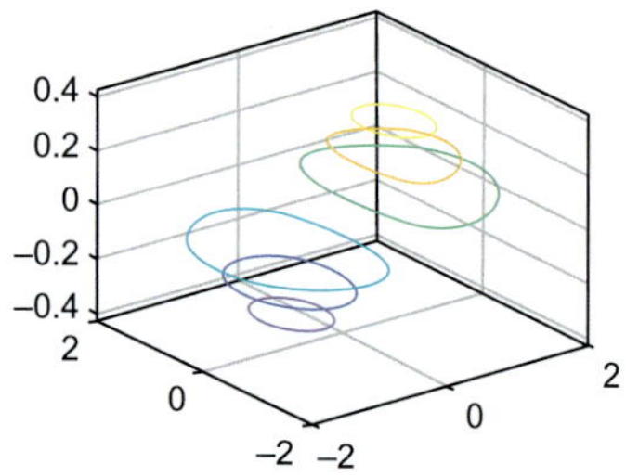

`cylinder` or `cylinder(r)` draws an ordinary or a profiled cylinder; the profile is given by the r expression	`>> t = 0:pi/10:2*pi;`
	`>> r=acot(t);`
	`>> cylinder(r)`
		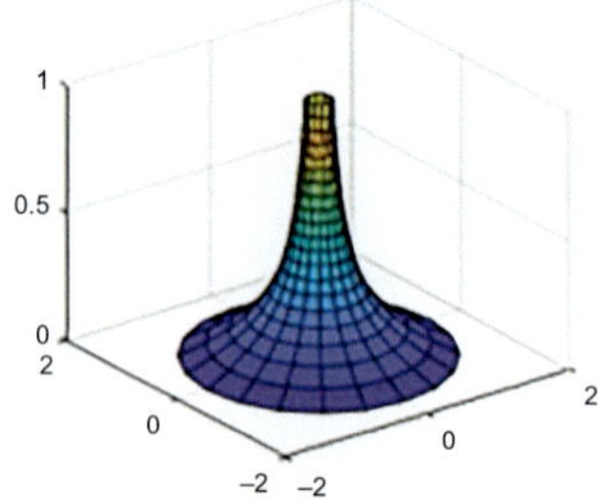

`figure` generates a new figure window named as Fig. 1; `figure(n)` generates a Figure Window with number n in the figure name	`>>figure(3)`	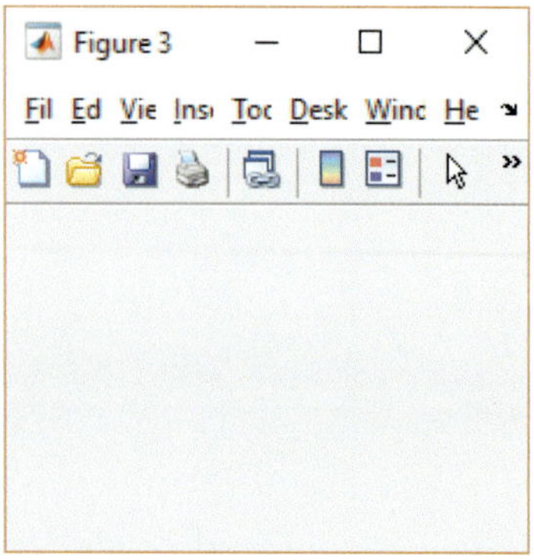

Table 3.5 Supplementary 2D and 3D plotting commands—cont'd

Commands and their purposes	MATLAB examples	Generated plots
`fplot('function',limits)` plots a function $y=f(x)$ with specified `limits` for x-axis (the limits of y-axis may be added); `'function'` is the function written as string	`>> fplot('2*exp(.5*x)',[0,10])`	

`loglog(x,y)` generates a plot with a log scaled (base 10) on both the x- and the y-axes	`>> x=linspace(0.05,30,200);`
	`>> y=4.7+exp(-1./x.^2);`
	`>> loglog(x,y),grid on`	

`pie(x)` generates a pie-like chart by the x data. Each element in x is represented as a pie slice	`>> x=[55 59 80 77 98 90];`
	`>> pie(x)`
	`>> title('Group Grades')`	

`pie3(x,explode)` generates a pie chart; `explode` is a vector that specifies an offset of a slice from the center of the chart; 1 denotes an offseted slice and 0 a plain slice	`>> x=[55 59 80 77 98 90];`
	`>> explode=[0 0 1 1 0 0];`
	`>> pie3(x,explode)`
	`>> title('Group Grades')`	

`polar(theta,rho)` generates a plot in polar coordinates in which `theta` and `rho` are the angle and radius respectively	`>> t=linspace(0,2*pi);`
	`>> ro=sin(1/3*t).*cos(1/3*t);`
	`>> polar(t,ro)`	

Continued

Table 3.5 Supplementary 2D and 3D plotting commands—cont'd

Commands and their purposes	MATLAB examples	Generated plots
`sphere` or `sphere(n)` plots a sphere with 20x20 (default) or nxn mesh cells respectively	`>> sphere(30), axis equal`	

`stem(x,y)` displays data as stems extending from a baseline along the x-axis. A circle (default) terminates each stem	`>> x=10:10:100;`
	`>> y=[1,2,1,2,5,2,5,2,3,1];`
	`>> stem(x,y)`
	`>> xlabel('Grades')`
	`>> ylabel('Number of Grades per Bin')`	

`stem3(x,y,z)` generates a 3D plot with stems from the x, y-plane to the z-values. A circle (default) terminates each stem	`>> x=0:0.25:1;`
	`>> [X,Y]=meshgrid(x);`
	`>> z=(X.*Y./(X+Y)).^1.2;`
	`>> stem3(X,Y,z)`	

`stairs(x,y)` generates a stairs-like plot of the discrete y data given at the specified x points	`>> x=2000:5:2025;`
	`>> y=[26 28 30 31 37 46];`
	`>> stairs(x,y)`
	`>> xlabel('Year')`
	`>> ylabel('People, mln')`	

`surfc(X,Y,Z)` generates surface and contour plots together	`>> x=-2:.2:2;`
	`>> [X,Y] = meshgrid(x);`
	`>> Z=2e3*X.*Y.*exp(-X.^2-Y.^2);`
	`>> surfc(Z);`	

Based on Table 3.3 Burstein, L. (2015). *MATLAB in Quality Assurance Sciences*. Cambridge: Elsevier-WP.

3.4 Application examples

3.4.1 Surface tension of fluid as a function of temperature

Surface tension γ of fluids is changed with temperature T. For ordinal water, the surface tensions are: γ=74.23 72.74 71.19 69.59 67.93 66.24 64.47 62.68 60.82, and 58.92 mN/m with the uncertainties equal to ±0.5% of γ, mN/m. These values are measured at temperatures t= 10°C, 20°C, 30°C, 40°C, 50°C, 60°C, 70°C, 80°C, 90°C, and 100°C. Empirical expression for the data is:

$$\gamma = 235.8\left(1 - \frac{T}{T_c}\right)^{1.256}\left[1 - 0.625\left(1 - \frac{T}{T_c}\right)\right]$$

where T_c equal to 647.096 K, $T = 273.15 + t$, K.

Problem: Represent this experimental data with the error in the plot with bars showing the error limits at each point. Add the empirical data calculated with the presented expression to the plot.

The commands to the problem solution are:

```
>>G_exper=[74.23 72.74 71.19 69.59 67.93 66.24 64.47 62.68 60.82 58.92];
>>t=273.15+[10:10:100];                    % vector of temperatures in K
>>Tc=647.096;
>>error_w=0.005;                           % error in parts of percent
>>G_theor=235.8*(1-t./Tc).^1.256.*(1-0.625*(1-t./Tc));
>>e=error_w*G_exper;                       % absolute surface tension errors
>>figure
>>errorbar(t,G_exper,e,'o')                % plots experimental points
>>hold on
>>plot(t,G_theor)                          % plots calculated empirical curve
>>xlabel('Temperature, K'),ylabel('Surface tension, mN/m')
>>title('Water surface tension')
>>legend('experemental points','theoretical curve')
>>grid on
>>axis tight                               % sets axes to data limits
>>hold off
```

The resulting plot is:

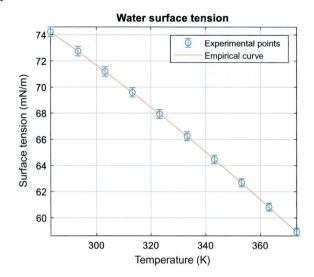

3.4.2 Stress-strain graph generated by the experimental data

In the tension testing of a material, the following data were obtained: load $F = 0, 703, 1334, 1906, 2568, 3404, 4003, 4701, 5338, 6041, 6672, 7084, 7384, 7326, 7117, 6760$, and $6316\,N$, and appropriate sample lengths $l = 50, 50.1865, 50.274, 50.374, 50.5005, 50.5935, 50.7005, 50.8, 50.9501, 51.15, 51.4005, 51.75, 52.1005, 52.35, 52.675$, and $53\,mm$. Diameter of the tested sample is $d = 10\,mm$. Values of the stresses σ and strains ε may be calculated by the expressions $\sigma = F/[(\pi d^2)/4]$ and $\varepsilon = (l - L)/L$ where l and L are lengths under and before loading respectively.

Problem: Calculate the stress and strain values by the given load—length data and plot the stress-strain graph.

The commands to solve this problem are:

```
>>F=[0 703 1334 1906 2568 3404 4003 4701 5338 6041 6672 7084 7384 7326 7117 6760 6316];
>>l=[50 50.1002 50.1865 50.274 50.374 50.5005 50.5935 50.7005 50.8 50.9501 51.15 51.4005 51.75 52.1005 52.35 52.675 53];
>>L=l(1);                                          % sample length
>>d=10;                                            % sample diameter
>>A=pi*d^2/4;                                      % sample section area
>>stress=F/A;                                      %stress calculation
>>strain=(l-L)./L;                                 %element-wise strain calculation
>>plot(strain,stress,'-o')                         %plots stress-strain graph
>>xlabel('Strain, ndm'),ylabel('Stress, N/mm^2')
>>title('Stress-strain curve')
>>grid on
```

The resulting plot is as follows:

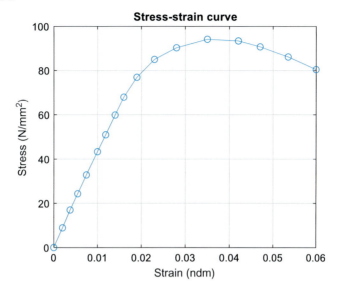

3.4.3 The Lennard-Jones interatomic potential

The interaction energy u between pairs of atoms or molecules, changing with interparticle distance r is described with the so-termed 12-6 Lennard-Jones potential, which in dimensionless form reads:

$$U = 4\left(\frac{1}{R^{12}} - \frac{1}{R^6}\right)$$

where R is reduced distance (dimensional distance divided by the distance σ at zero potential energy), and U is reduced interatomic energy (dimensional energy divided by the depth ε of the potential well).

Problem: calculate and plot the potential energy U as function of distance at R between 0.9 and 2.

The commands that should be entered to solve this problem:

```
>>R=linspace(0.9,2);          % sets 100 values between 0.9 ... 2
>>U=4*(1./R.^12-1./R.^6);     % element-wise U calculations
>>plot(R,U)                   % plots U,R -graph
>>xlabel('Distance, r/\sigma'),ylabel('Potential energy, u/\epsilon')
>>title('The 12-6 Lennard-Jones potential')
>>grid on
>>axis tight                  % sets axes to the data limits
```

Generated plot is:

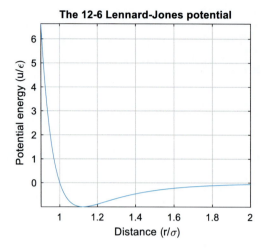

3.4.4 Transient one-dimensional diffusion

The approximate form of a 1D analytical solution of the diffusion equation is:

$$c = \frac{M}{\sqrt{4\pi D t}} e^{-\frac{x^2}{4Dt}}$$

where c is the concentration of a substance (in g/cm^3) that changes with coordinate x and time t (in s) due to the diffusion process; M is the substance mass per area (in g/cm^2), D is the diffusion coefficient (in cm^2/s).

Problem: Calculate c and generate a three-dimensional surface plot in which x and t are the horizontal plane axes, and the z-axis is the concentration; assume D equal to 1.64 cm^2/s (gaseous H$_2$), $M=1$ g/cm^2, $x=0$ ... 1 cm (make 10 equally spaced values), and $t=0.1$ ... 2.1 s (make 20 equally spaced values). Present the plot with the azimuth and elevation angles 100 and 25 degrees respectively.

The program follows these steps:

– Create a 10-element vector with x-values and 20-element vector with t-values; set the D and M-values;
– Create an X,Y-grid in the ranges of the x and t vectors, respectively, by using for this the `meshgrid` command;
– Calculate c for each pair of X and Y values by the above expression applying the element-wise operations;
– Generate a 3D plot by the determined X,Y,c-values;
– Set the required view point for the generated plot.

The commands that actualizes this steps:

```
>>x=linspace(0,1,10);              % vector x with 10 values, cm
>>t=linspace(0.1,2.1,20);          % vector t with 20 values, sec
>>D=1.64;                          % cm^2/sec
>>M=1;                             % kg/cm^2
>> [X,Y]=meshgrid(x,t);            % creates X,Y grid from the x,t vectors
>>c=M./sqrt(4*pi*D*Y).*exp(-X.^2./(4*D*Y)); % calculate concentration
>>surf(X,Y,c);                     % plot surface graph
>>xlabel('Distance, cm');ylabel('Time, sec');zlabel('Concentration, g/cm^3')
>>view(100,25)                     % az=100° el=25°
```

The generated plot is:

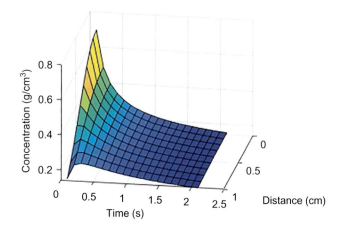

3.4.5 Temperature of a square plate

Two-dimensional temperature distribution $T(x,y)$, °C, in the a x a square plate with temperature T_1 on an one of the boundaries and zero-temperatures on the other boundaries is described by the expression:

$$T = \frac{4T_1}{\pi} \sum_{i=1,3,5,\ldots}^{n} \frac{\sinh(i\pi x/a)\sin(i\pi y/a)}{i\sinh(i\pi)}$$

Problem: Generate $T(x,y)$ plot by the expression when x and y are equal to 0 ... a (make 20 equally spaced x and y values each), $a=2$ m, $T_1=75$ °C, $n=51$. Present the plot with the azimuth and elevation angles of 20 degrees each.

To solve this problem, we follow these steps:

– Set a, T_1 and n values;
– Create the 20-element vector with x-values between 0 and a;
– Create X,Y-grid in the ranges of the x vector by using for the `meshgrid` command;

112 A MATLAB® Primer for Technical Programming in Materials Science and Engineering

- Calculate and save *T* for each pair of *X* and *Y* values with two loops; the loops are constructed such that for each *X* all *Y* values are passed, and into each loop pass the `sum` command is realized the series summating (in last operation the element-wise operations should be used);
- Generate a 3D plot by the determined *X,Y,T*-values;
- Set required view point for the generated plot.

The commands are:

```
>> a=2;
>> T1=75;
>> n=51;
>> x=linspace(0,a,20);                    % vector x with 20 values
>> y=x;                                   % vector y with 20 values
>> [X,Y]=meshgrid(x,y);                   % sets X and Y coordinate matrices
>> for i_x=1:length(x)                    % loop for x-coordinates
     for j_y=1:length(y)                  % loop for y-coordinates
       i=1:2:n;                           % vector of the series term numbers
T(j_y,i_x)=4*T1/pi*sum(sinh(i*pi*X(i_x,j_y)/a).*sin(i*pi*Y(i_x,j_y)/a)./(i.*sinh(i*pi)));
     end
   end
>> surf(x,y,T)                            % 3D plot
>> xlabel('x, m'),ylabel('y, m'),zlabel('T,^oC')
>> title('Square plate temperature distribution')
>> view(20,20)                            % az=20° el=20
```

The generated plot is:

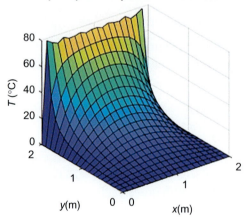

3.4.6 Velocity distribution of the gas molecules

Molecules of gas have different velocities that change with temperature. The distribution function f_v of the molecules velocity v is described by the Maxwell-Boltzmann equation:

$$f_v = 4\pi v^2 \left(\frac{M}{2\pi RT}\right)^{\frac{3}{2}} v^2 e^{-\frac{Mv^2}{2RT}}$$

where M is the molecular weight, T is temperature, and R is gas constant.

Problem: Generate the $f_v(v,T)$ plot when v and T are in the ranges 0–1500 m/s and 50–500 K, respectively; $M = 0.004$ kg/mol., and $R = 8.314$ J/(K mol.); make 20 equally spaced values of v and of T each. Present the plot with the azimuth and elevation angles of 21 and 19 degrees respectively.

The steps for the problem solution are:

- Assign M and R values;
- Create a 20-element vector of the v-values between 0 and 1300;
- Create a 20-element vector of the T-values between 50 and 500;
- Create an X,Y-grid in the ranges of the v and T vectors, respectively, by the `meshgrid` command;
- Calculate f_v for each pair of X and Y values; the element-wise operations should be used;
- Generate a 3D plot by the determined X,Y,f_v-values.

The commands solving this problem are:

```
>>M=4e-3;
>>R=8.314;
>>v=linspace(0,1500,20);              % vector v, 20 values
>>T= linspace(50,500,20);             % vector T, 20 values
>>[X,Y]=meshgrid(v,T);        % matrices X and Y of the v and T values
>>fv=4*pi*(M./(2*pi*R*Y)).^(3/2).*X.^2.*exp(-M.*X.^2./(2*R*Y));
>>surf(X,Y,fv)                        % plots surface graph
>>xlabel('Velocity, m/s'),ylabel('Temperature, K'),zlabel('Disribution function')
>>title('Maxwell-Boltzmann distribution')
>>grid on
>>view(21,19)
```

The resulting plot is:

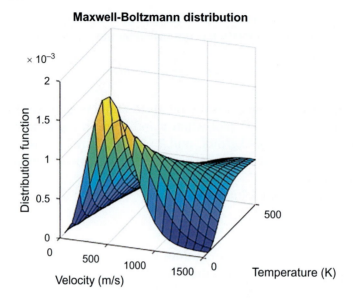

3.5 Questions and exercises for self-testing

1. To designate a point in a plot with an inverted triangle, the following word/sign/character should be used: (a) "triangle," (b) +, (c) ^, (d) v. Choose all the correct answers.
2. The plot3 command generates a plot showing: (a) a 3D surface, (b) a 3D mesh, (c) a 3D line, (d) a 2D grid. Choose all the correct answers.
3. To divide page (i.e., Figure window) to place several plots in the same page, it is necessary to enter the following command: (a) hold on, (b) subplot, (c) plot(x1, y1,x2,y2, ...), (d) surf. Choose all the correct answers.
4. To remove axes from a plot, it is necessary to use the command: (a) hold off, (b) delete('axes'), (c) grid off, (d) axis off. Choose all the correct answers.
5. The hist command draws bars with heights that are equal to: (a) frequencies of data, (b) data values, (c) possibly both options (a) and (b), (d) number of data points in the range equal to the bar width. Choose all the correct answers.
6. A 2D graph without a line between the data points can be produced using the command: (a) plot(x,y,'o'), (b) scatter(x,y), (c) both options (a) and (b) are correct, (d) there is no correct answer. Choose the correct answer.
7. The exponential function $y = e^{3.1x}$ can be drawn as straight line with the command: (a) semilogx, (b) semilogy, (c) loglog, (d) there is no correct answer. Choose all the correct answers.
8. To generate a plot with the necessary point of view, the following should be done: (a) rotate the plot with the mouse, (b) write and input view command with the

appropriate azimuth and elevation angles, (c) both options (a) and (b) are correct, (d) there is no correct answer. Choose all the correct answers.

9. Two different y-axes in the same 2D plot can be activated with the following command/s should be used: (a) `yyaxis left`, (b) `yyaxis right`, (c) both the `yyaxis left` and `yyaxis right`, (d) there is no correct answer. Choose all the correct answers.

10. In a 2D plot, the line color can be changed with: (a) the `colormap` command, (b) the appropriate specifiers into the `plot` command, (c) both options (a) and (b) are correct, (d) there is no correct answer. Choose all the correct answers.

11. An applied material company tested the thermal conductivity λ of a commercial metal at low temperatures T. The λ-values are 423, 385, 358, 351, 346, 347, 350, and 355 W/(m K) at corresponding T = 80.06, 95.34, 115.62, 135.53, 159.46, 181.56, 198.35, and 217.30 K. Plot the data with solid line between the points designated with diamond, add grid, axis labels, and a caption.

12. The kinematic viscosity η, mm²/s, as function of temperature T, °C, of a motor oil was roughly fitted with the Reynolds-type equation:

$$\eta = Ae^{BT}$$

where A = 906.8, B = −0.0504.

The measured η-values are 658, 327, 178, 105, 66, and 45 31 mm²/s at T = 10°C, 20°C, 30°C, 40°C, 50°C, and 60°C respectively. The two-sided uncertainties (i.e., errors) in the kinematic values are 0.02 η mm²/s. Generate two plots on one page (i.e., in the same Figure window), the left plot with the logarithmic y-coordinate and measured kinematic viscosity, and the right plot in ordinary x,y-coordinates with both calculated and measured viscosity values; give this plot the bar of errors at each measured point and add legend; in each plot, designate the measured points with circle and add grid, axis labels, and captions.

13. The measured surface tension and density of the benzene as function of temperature are presented in the table.

T (K)	293.15	298.15	303.15	308.15	313.15	318.15	323.15
γ (mN/m)	29.22	28.52	27.81	27.14	26.45	–	–
ρ (g/cm³)	0.8789	0.8736	0.8683	0.8629	0.8576	0.8522	0.8467

Represent these data in the plot with two y-axes, add grid, axis labels, a caption and legend to the plot; mark the surface tension and density data points with the triangle and inverted triangle signs respectively; set dashed and solid styles for lines between the surface tension and density points respectively.

14. The solid-liquid equilibrium (melting line) temperatures T for various mole fractions of nitrogen x_{N_2} mixed with methane are measured in two separate experiments. The first T and x_{N_2} experimental data is 63.3, 63.4, 63.10, 65.25, 68.60, 71.70, 75.10,

78.60, 82.50, 86.45, 90.65 K, while x_{N_2} = 1.000, 0.919, 0.804, 0.708, 0.602, 0.504, 0.399, 0.300, 0.199, 0.104, 0.000 mol./mol. respectively. The second set of T_2 and x_{N_2} data are 79.05, 77.20, 75.20, 74.20, 70.80 K and 0.302, 0.363, 0.417, 0.453, 0.549 mol./mol., respectively.

Generate the $T(x)$ plot; designate the first data points with green triangles and the second with the blue inverted triangles. Connect the points of the first data set with the dashed line; do not connect the second set points. Add a grid, axis labels, caption, and legend to the plot.

15. The phase transition temperatures T versus weight w data of the binary alloy (e.g., Cu-Ni-type) is T_1 = 1083°C, 1190°C, 1280°C, 1360°C, 1420°C, and 1452°C while T_2 = 1083°C, 1120°C, 1170°C, 1250°C, 1340°C, and 1452°C. Both component temperatures are given at the various weight fractions of the second component w = 0, 20, 40, 60, 80, 100 Wt % comp2 (weight percentage of the second component).

Generate a phase diagram by the data points with the solid line between them; mark the first component points with the green circles and the second—with the black squares. Add the "Fluid," "Solid," and "F+S" text descriptions in the following (x,y)-positions—(20,1330), (70,1170), and (42,1250), respectively. Add a grid, axis labels, caption, and legend to the plot.

16. The van der Waals equation of state is a rather accurate approach for a material in gaseous phase:

$$p = \frac{nRT}{V-b} - \frac{n^2 a}{V^2}$$

where p is the pressure, atm; V is volume, L; T is temperature, K; n is the mole amount; R is gas constant, 0.08206 L atm/(mol. K), a and b are material constants in the $L^2 \cdot$atm/mol.^{2} and L/mol. units, respectively.

Generate $p(T,V)$ plot for 1.5 mol of gas when a = 1.36 L^2 atm/mol.^{2}, b = 0.03183 L/mol. (oxygen) and T and V are given on 20 equidistantly spaced points (for each variable) in the range 273.15–523.15 K and 0.1–2 L, respectively. Add axis labels and a caption to the plot.

17. Reaction rates in accordance with the material kinetics (e.g., in polymeric materials producing) are described by the Arrhenius equation:

$$k = A e^{-\frac{E_a}{RT}}$$

where k is the reaction rate, E_a is activation energy, T is temperature, A is prefactor, R is gas constant.

Generate $k(E_a, T)$ plot when A = 10^{14} s^{-1}, R = 8.314 J/(mol. K), and E_a and T are in the ranges 19000 … 25000 J/mol. and 293 … 303 K respectively; take 20 equally spaced E_a and T values. Add axis labels and a caption to the plot. Make the axis limits equal to the data limits.

18. The boson energy distribution function, in accordance with the Bose-Einstein statistics, is:

$$f_E = \frac{1}{e^{\frac{E}{kT}} - 1}$$

where E is energy, eV; T is temperature, K; k is the Boltzmann' constant equal to $8.617 \cdot 10^{-5}$ eV/K.

Generate the $f_E(E,T)$ plot for values E and T in the range 1 ... 2 eV and 1000 ... 5000 K respectively. Assume 20 values of E and T each.

19. In the Langmuir adsorption model, the solid surface coverage θ is function of an equilibrium pressure P and the Langmuir coefficient b:

$$\theta = \frac{bP}{1 + bP}$$

where b is the constant given for a substance at a certain temperature.

Build two plots in two separate Figure windows (using the two `figure` commands):
- the $\theta(P,b)$ graph for P and b are in the ranges 0, ..., 3 (make 50 equally spaced values) and 0.2, ..., 50 (make 25 equally spaced values) respectively;
- the $\theta(P)$ graph for three b constants—5, 15, and 50.

Add axis labels, a grid, and caption to each plot, and add a legend to the $\theta(P)$ plot.

20. Generate four plots on the one page with the following bodies:

(a) Cross-cap

$$x = \cos u \cdot \sin 2v$$
$$y = \sin u \cdot \sin 2v$$
$$z = \cos^2 v - \cos^2 u \cdot \sin^2 v$$

where u and v range between 0 and 2π; make 40 equally spaced values of u and v each.

(b) Möbius' strip

$$x = \left(1 + \frac{1}{2}v\cos\frac{u}{2}\right)\cos u$$
$$y = \left(1 + \frac{1}{2}v\cos\frac{u}{2}\right)\sin u$$
$$z = \frac{1}{2}v\sin\frac{u}{2}$$

where u and v ranges are 0 ... 2π and -1 ... 1 respectively; make 20 equally spaced values of u and v each.

(c) Toroid

$$x = (R + r\cos v)\cos u$$
$$y = (R + r\cos v)\sin u$$
$$z = r\sin v$$

where $R=6$, $r=2$, u and v ranges are $0 \ldots 2\pi$ each; make 20 equally spaced values of u and v each.

(d) Ellipsoid

$$x = a\cos u \cdot \sin v$$
$$y = b\sin u \cdot \sin v$$
$$z = c\cos v$$

where $a=1.2$, $b=1.7$, $c=2.1$, $u=0, \ldots, 2\pi$, $v=0, \ldots, \pi$; make 20 equally spaced values of u and v each.

Remove axes and add captions to each plot.

21. The stress field due to so-called edge dislocation is given by the three following equations:

$$\sigma_{xx} = -\frac{Gb}{2\pi(1-v)} \frac{y(3x^2+y^2)}{(x^2+y^2)^2}$$

$$\sigma_{yy} = \frac{Gb}{2\pi(1-v)} \frac{y(x^2-y^2)}{(x^2+y^2)^2}$$

$$\tau_{xy} = \frac{Gb}{2\pi(1-v)} \frac{x(x^2-y^2)}{(x^2+y^2)^2}$$

Generate on the one page three plots $\sigma_{xx}(x,y)$, $\sigma_{yy}(x,y)$ and $\tau_{xy}(x,y)$ as follows: the $\sigma_{xx}(x,y)$-plot in the left top pane; the $\sigma_{yy}(x,y)$-plot in the left bottom pane, and the $\tau_{xy}(x,y)$-plot in the two joined right panes. The x range is $-5 \cdot 10^{-9} \ldots 5 \cdot 10^{-9}$ m, y range is $-5 \cdot 10^{-9} \ldots -1 \cdot 10^{-9}$ m, $G=79.3 \cdot 10^9$ Pa, $b=0.248$ m, and $v=0.27$ dimensionless (steel). Add axis labels, a grid, and caption to each plot.

3.6 Answer to selected questions and exercises

2.
   (c) a 3D line.

4.
   (d) `axis off`.

6.
   (c) both options (a) and (b) are correct.

8.
   (b) write and input `view` command with the appropriate azimuth and elevation angles.

10.
   (b) the appropriate specifiers into the `plot` command.
12.

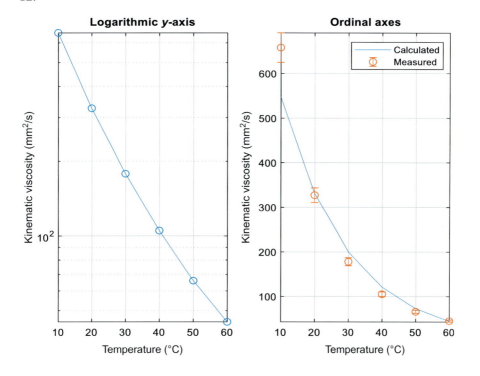

14.

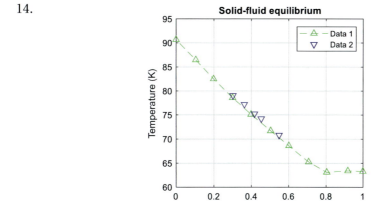

16.

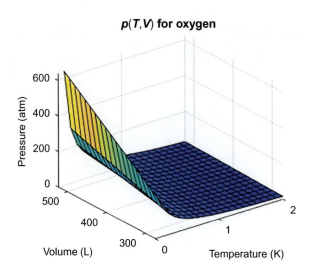

18.

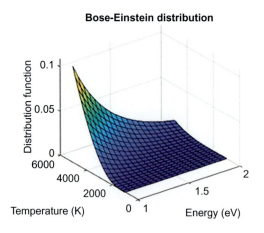

20.

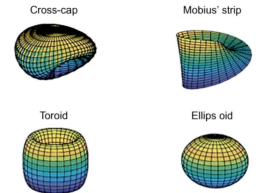

4

Writing programs for technical computing

In preceding chapters, all studied commands and manipulations with them were performed interactively; the commands were not stored and should be re-entered each time the calculations must be repeated. This is a disadvantage of the interactive mode in the Command Window; another disadvantage of this mode is that correction of the command is possible only on the executed line. When any of the series commands should be corrected, all predecessors together with this and subsequent commands must be repeated for obtaining the correct result. All this is inconvenient, and the reader who has learnt the preceding chapters attentively has undoubtedly experienced it. The remedy lies in writing a list of the commands (termed program), saving them into a file, and running it as necessary. There are two types of such files in MATLAB®: script and function files. These file types are explained further. In addition, some functions of numerical analysis frequently used in technical calculations in general and in materials science and technology in particularly, are presented in the chapter.

4.1 Scripts and script files

4.1.1 How to create, save, and run the script file

Sequence of the written commands is called script and is a program; the commands are executed sequentially in the order they were written. The script should be typed in the MATLAB® Editor window, saved in a script file, and run thereafter. Corrections and new commands can be entered directly in the file. Saved files have the extension ".m" and termed m-files.

To type script, first open the Editor; to do this, type and enter the `edit` command in the Command Window, or click the New Script button of the File group in the Desktop Home tab. After these operations are performed, the Editor window appears; in Fig. 4.1, the window is represented in its undocked (from the Desktop) view. The window contains toolstrip and blank fields for commands. The toolstrip is organized likely to the Desktop (see Section 2.1.1) and includes the EDITOR, PUBLISH, and VIEW tabs. The EDITOR tab (default appearance) is briefly described below as commonly used for writing/editing commands, saving/opening them, running, and debugging.

The script commands should be typed line by line in the blank field of the Editor window beginning from the place showing on the Fig. 4.1. Two or more commands can be typed on the same line with separating commas or semicolons. After pressing the Enter

124 A MATLAB® Primer for Technical Programming in Materials Science and Engineering

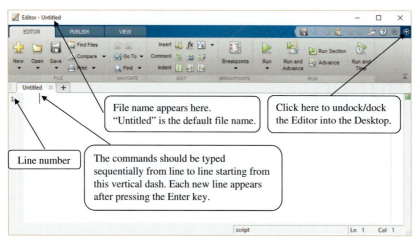

FIG. 4.1 The Editor window.

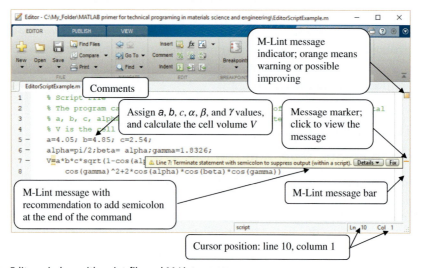

FIG. 4.2 The Editor window with script file and M-Lint message.

key, a new line will be accessible. Each opened line automatically gets its own number. After writing commands, they should be saved in a file with a desirable name. The commands can also be typed in any non-MATLAB® text editor; after that, they should be copied into the MATLAB® Editor window.

The Editor window with a typical script file is shown in Fig. 4.2. The file is named EditorScriptExample and calculates triclinic unit cell volume; its first four lines are explanatory comments, preceded by the comments sign % and in green; they are not part of the execution. The four further lines content commands designed to calculate and display the cell volume V by the assigned a, b, c, α, β, and γ values (see the same calculation in interactive mode—Section 2.1.8.1); the commands in the Editor window appear in black for better legibility.

The bar in the right vertical frame of the Editor Window is a place when the message markers (horizontal colored dashes) of the code analyzer appear, also called M-Lint analyzer, which detects possible errors, comments on them, and recommends corrections for better program performance. The square sign ▢ at the top of the message bar indicates the presence or absence of errors and/or makes some warnings. When the indicator is green, it means no errors, warnings, or possibility of improvements; a red indicator—syntax errors detected; an orange indicator—warnings or possibility of improvements (but no errors). When the analyzer detects an error or possibility of improvement, it underlines/highlights the text/character and horizontal colored line appears on the message bar. Moving the cursor to this line or to the highlighted place, we can obtain the comment message—see Fig. 4.2. Not every comment should be considered; for example, the recommendation shown in this figure (addition of a semicolon) need not be executed because we want to display the resulting value of the cell volume. The code analyzer regime is default mode of the Editor window; it can be disabled by un-signing the "Enable integrated warning and error messages" check box in the "Code Analyzer" option of the Preferences window that can be opened by clicking the ⚙ Preferences line in the Environment section of the MATLAB® Desktop Home tab.

4.1.1.1 Saving the script file

After the commands have been written into the Editor window, they should be saved as a file. For this, the "Save As …" line of the Save button (from the FILE group of the EDITOR tab) should be chosen; the "Select File for Save As" window appears, both the desired file location and a name for file should be typed respectively into the path- and "File name:" fields of this window. The file saves into the "MATLAB®" folder of the "My documents" directory by default; nevertheless, any other directory can be chosen from the tree which appears on the left-hand side of the "Save File for Save As" window.

When naming a file, observe the following rules:

- the name should begin with a letter and cannot be longer than 63 characters;
- the name should not repeat names of the MATLAB® commands/functions, user-defined functions, or predefined variables;
- the mathematical operation signs (e.g., +, -, /, \, *, ^) should not be used within the file name;
- it is strongly recommended to not introduce spaces into the file name.

4.1.1.2 About the Current folder

The directories and files of the currently used folder are shown in the Current Folder window located to the left of the Desktop. Additionally, the path to the used folder is shown in the current folder field of the toolbar strip located just under the Desktop toolstrip—see Fig. 4.3. To set the Command Window to a nondefault folder, e.g., to the folder with m-file previously created and saved with Editor, the following operations should be performed:

126 A MATLAB® Primer for Technical Programming in Materials Science and Engineering

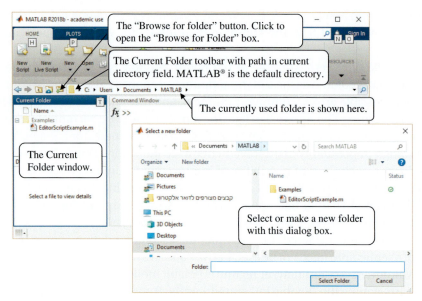

FIG. 4.3 The "Current Folder" window, current folder toolbar, "Browse for folder" button, and opened "Select a new folder" dialog box.

- click the icon ![icon] located on the left-hand side of the current directory field, after that the "Select a new folder" dialog box appears;
- select and click on line with desired directory and then click the "Select Folder" button; selected directory appears in the current directory field. The current Desktop folder also appears in the Editor window.

For example, the Fig. 4.2 shows the EditorScriptFile.m stored next to the Examples subfolder with the following path to: C:\Users\Documents\MATLAB®.

4.1.1.3 Running script file
For executing a script file:

- the available files in the current MATLAB® folder should be checked first, and if the file is not in this folder, an appropriate directory containing this file should be set, by the way described above;
- the file name (without m-extension) should be typed in the Command Window and then entered.

Following this, to run the `EditorScriptExample.m`, we need to change the current directory to the C:\Users\Documents\MATLAB® by the described way, and then to type and enter the file name in the Command Window:

```
>> EditorScriptExample        % enter to run the script
V =
48.1919
```

The unit cell volume V calculated by the commands saved in the script file named `EditorScriptExample`.

4.1.2 Input the variable values from the Command Window

To use the script file for recalculations with new parameters, it is necessary to type new values into the script file, resave, and run it. It is inconvenient to change the script for each new parametrical calculations. To avoid writing values strictly into the script file, the `input` command should be used; this command has forms:

```
Numeric_Variable = input('Text displayed in the Command Window')

Character_Variable = input('Text displayed in the Command Window','s')
```

where `'Text displayed in the Command Window'` is a message that displays in the Command Window and enables to assign: a number to the `Numeric_Variable`, or a string to the `Character_Variable`; `'s'` signs that inputting text is a string, and it can be written without single quotes. For a string containing more than one line, the `\n` specifier should be entered at the end of the line.

After running a script file, when the `input` command is initiated, the text written in this command is displayed on the screen and the user should type and enter a number or string; inputted values are assigned to the `Numeric_Variable` or to the `Character_Variable` depending on the used command form.

As an example for using this command: create script file that convert US customary pressure units, psi—pounds per square inch, into international system SI standard units, MPa—mega pascals, by the expression $P_{Pa} = 0.00689476 \cdot P_{psi}$:

```
% psi to Pa convertor
psi=input('Enter pressure in psi, p = ');
P_Pa=6894.76e-6*P_psi;
fprintf('\n Pressure in MPa is %10.4f\n',P_Pa)
```

The commands should be typed into the Editor window and saved in the m-file; name the file as `psi2Pa`. After entering this file name in the Command Window, the prompt "Enter pressure in psi, p =" appears on the screen; type a pressure value (in psi) and press enter; the psi pressure value is converted to Pa and the "Pressure in MPa is" string together with the obtained value displayed on the screen.

Running command, appeared prompts, inputted pressure in pounds per square inch, displayed string, and defined pressure in pascals are:

```
>> psi2Pa
Enter pressure in psi, p = 14.7
Pressure in MPa is 0.10135
```

The `input` command enables inputting the vectors and matrices; this can be done by printing the numbers into the brackets. Use, for example, the psi2Pa script file to convert two pressure values:

```
>> psi2Pa
Enter pressure in psi, p = [14.7 1.47e3]
```

```
Pressure in MPa is 0.10135
Pressure in MPa is 10.13530
```

4.2 User-defined functions and function files

4.2.1 Creating the user-defined function

In mathematics, a function f relates an inputted, e.g., x, and an outputted, e.g., y, parameters (arguments), and the simplest form of a function is $y=f(x)$. In the right and left parts can be sets of parameters, e.g., $\mathbf{u}=f(x_1,x_2,x_3)$ with $\mathbf{u}$ as a three-element vector. When $f(x)$ is specified as an expression, and the inputted parameters are assigned, the outputted parameters can be obtained. Many commands discussed in the preceding chapters were written in the function form, e.g., sqrt(x), sin(x), cos(x), exp(x), log(x), etc.; this form enables to use the commands in simple direct calculations or in complicated expressions by typing function name with the appropriate argument. In addition to existing functions, MATLAB® provides possibility to create any new function and then reuse it with arbitrary argument values and in different programs; such functions are called "user-defined." Not only a specific expression, but a complete program created by the user can be defined as a function and saved as a function file. This is especially useful in cases where there is no desirable built-in function for proper calculation.

In general, the user-defined function should contain the following parts: line with function definition, help lines with explanations, and a body with program commands. All these components should be written in the Editor window. An example of a user-defined function is presented in Fig. 4.4.

The function and its file is named Far2Cels and converts the temperatures given in degrees Fahrenheit to the degrees Celsius and Kelvin (SI units). According to the first definition line, the function has: one input parameter—the vector t_F with temperatures in

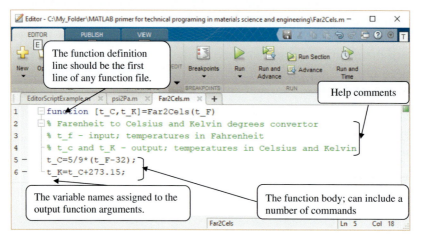

FIG. 4.4 Typical view of the functions file containing a user-defined function.

Fahrenheit that should be converted and two output parameters—the converted temperatures in Celsius and Kelvin.

The requirements and recommendations regarding each part of the function file are as follows.

4.2.1.1 Definition line of the function

The line with definition of a function reads:

```
function [output _parameters]=function_name(input_ parameters)
```

The word `function` must be the first word of the user-defined function and appears in blue. The `function_name` is a name that is written to the right of the " = " sign and should obey the same rules as for variable names (Section 2.1.4). The `input_parameters` arguments are a list of variables which values should be transferred into the function; the `output _ parameters` are a list of those we want to process and derive from the function. The input parameters must be written between the parentheses and the output parameters—between square brackets. In the case of a single output parameter, the brackets can be dispensed with. When there is more than one input/output parameter, they should be divided by commas.

The input and output parameters of the function can be omitted completely or partially. Possible views of the function definition line with or without input/output parameters are given in the following examples:

`function [A,B]=example1(a,b,c)` – full record, function named `'example1'` with three input (a, b, and c) and two output (A and B) parameters;

`function D=example2(a,b)` – full record, function named `'example2'` with two input (a and b) and one output (D) parameters;

`function [A,B,C]=example3` – function named `'example3'` with three output parameters (A, B, and C), no input arguments;

`function example4(a,b,c)` – function named `'example4'` with three input arguments (a, b, and c), no output parameters;

`function example5` – the function named `'example5'` with no input and output parameters.

The amounts and names of the input and output arguments can differ from those in the examples. Note: the "function" word should be written in lower-case letters.

4.2.1.2 Lines with the help comments

Lines with the help comments should be placed just after the function definition line. The first help line should contain a short definition of the function. (This line is displayed by the `lookfor` command when it searches the information throughout all the MATLAB® paths.) All lines of the help comments are displayed when the `help` command is introduced with a name of the user-defined function, e.g., typing into the Command Window Command Window `help Far2Cels` yields: yields:

>> help Far2Cels
Farenheit to Celsius and Kelvin degrees convertor
t_f - input; temperatures in Fahrenheit
t_c and t_K - output; temperatures in Celsius and Kelvin

Note, the help comments are not obligatory for the user-defined function, and can be omitted.

Hereafter, to be short, the user-defined function help part will be written in short form—two lines only: line of function explanation and line with the run command example.

4.2.1.3 Function body, local, and global variables

The function body may contain one or more commands for actual calculations. Between these commands, frequently at their end, should be placed the assignments to the output parameters; e.g., in the example in Fig. 4.4, the output parameters are t_C and t_K; thus the last two commands calculate and assign defined values to the t_C and t_K. When we run the function, the actual values must be assigned to the input arguments: what is necessary to perform the calculations within the function body.

The function body can include another used-defined function/s that is/are called in this case the subfunction/s.

The variables in the function file are local and relevant only within commands of this function body. That means if the function has been run and finished, its variables are not saved and are no longer stayed in the workspace. If we want to share some or all of them with other function/s, we should make them accessible, which can be done with the `global` command having the form:

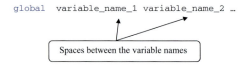

The `global` command with one or more variable names (designated here as `variable_name_1 variable_name_2 ...`) must be written before the variable/s is/are firstly used in the function and should be repeated also in other functions where it/s is/are intended to be used.

To display the value of a global variable in the Command Window, e.g., to view global variable *x*, previously defined in some user-defined function, the `>>global x` should be entered before typing `>> x` in the Command Window.

4.2.2 Function file

The function written in the Editor window should be saved in a file before running it. This is implemented exactly so as for a script file: select "Save As" line of the Save option from the File group of the Editor tab and then enter the desired location and name of the file. It is strongly recommended to call the file by the function name, e.g., the `Far2Cels` function should be saved in a file named `Far2Cels.m`.

Examples of function definition lines and corresponding names of the function files:

`function c=mydiffusion(m, d_coeff)`—the function file should be named and saved as `mydiffusion.m`;
`function [HRB, HRC]=hardness(F,D,di)`—the function file should be named and saved as `hardness.m`;
`function enthalpy(P,T)`—the function file should be named and saved as `enthalpy.m`.

The name of a function file containing two or more subfunctions should be identical to the name of the main function (from which the program starts).

4.2.3 Running a user-defined function

The user-defined function saved in file can be run from another file/program or from the Command Window as follows: the function file definition line should be typed without the word "`function`" and to the input parameters should be assigned their values. For example, the `Far2Cels` function file (Fig. 4.4) can be run with the following command:

```
>> [t_C,t_K]=Far2Cels([80 120])
t_C =
  26.6667 48.8889
t_K =
  299.8167 322.0389
```

Or alternatively, with preassignment of the input variables:

```
>> t_C=[80 120];
>> [t_C,t_K]=Far2Cels(t_C)
t_C =
  26.6667 48.8889
t_K =
  299.8167 322.0389
```

A user-defined function or its output parameters can be used in another mathematical expression or program. For example, to calculate kT energy with Boltzmann's constant $k = 8.6173 \cdot 10^{-5}$ eV/K and $T = 80$ F, we should convert temperature to degrees Kelvin with the `Far2Cels` function and then multiply T by k; the following commands should be typed in the Command Window:

```
>> k=8.6173e-5;t_F=80;
>> [t_C,t_K]=Far2Cels(t_F);     % convert from degrees Fahrenheit to Kelvin
>> kT_energy = k* t_K            % uses the t_K output parameter of the Far2Cels function
kT_energy =
   0.0258
```

4.2.3.1 Script and user-defined function files, comparison

Beginners usually find it difficult to understand the differences between script and function files. Indeed, most of actual problems can be solved with the ordinal script-type file, or with interactive mode—simply by inputting commands directly into the Command Window. For a better understanding of these two file forms, their similarities and differences are outlined below.

- Both are created using the Editor and saved with the extension m, as m-files.
- A function file must contain the function definition line as the first line; this feature is absent in a script file.
- The name of the function file should be the same as that of the function; this requirement does not make sense for the script file because the latter does not have a function definition line.
- Function files can receive/return data through the input/output parameters respectively; script file does not have this possibility, and values of its variables should be assigned directly within the file, or inputted with the `input` commands.
- Only function files can be used as functions in other user-defined functions, or simply in the Command Window.
- Only script files use the variables that have been defined in the workspace.
- Which file is preferable to create for resolving the problem should be solved by the user.

4.3 Selected MATLAB® functions and its applications in MSE

Material sciences studies, technologies, and laboratory practice involves a variety of wide-used math operations, such as interpolation, extrapolation, nonlinear algebraic equation solution, minimax, integration, and differentiation. The available MATLAB® functions that can be used for listed purposes are described below.

4.3.1 The interp1 function for interpolation and extrapolation

Data obtained by experiment or by tables are available at certain points, and it is required sometimes to estimate values between the observed points; this can be done by interpolation. When values laid outside the data points need to be evaluated, this can be done by extrapolation. For example, in some experimental technology, a cylindrical form preheated substance is cooled; the cooling temperatures at certain times were measured. The data is 700, 503, 402, and 360°C at times $t = 1, 3, 5$, and 9 s respectively. The graph with measured, interpolated, and extrapolated points is represented in Fig. 4.5.

Finding the temperature values at $t = 1.3$ and 7 s (values between original data points) is an interpolation problem, and finding them at $t = 0.5$ and 9.5 s (values outside data points)—extrapolation problem.

Both actions: inter- and extrapolation can be performed in MATLAB® with the `interp1` function. The available forms are:

`yi=interp1(x,y,xi,'method')` or `yi=interp1(x,y,xi,'method','extrap')`

The first form presented is used for interpolating data; the second one is for extrapolation or for simultaneously inter- and extrapolation.

The output parameter of these functions is `yi`—the defined interpolated/extrapolated value or vector of values.

The input parameters are:

`x` and `y` are the vectors with data argument and function values respectively;

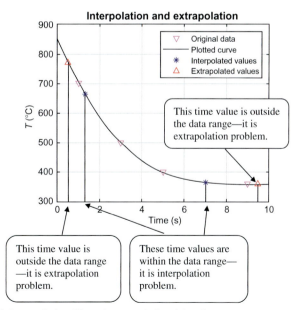

FIG. 4.5 Original data (∇), interpolation (*), and extrapolation (Δ) points.

`xi`—a scalar or a vector with the *x*-points (arguments) for which the values of `yi` (*y*-points) are sought;

`'method'` is the string containing name of the mathematical method to be used for inter- or extrapolation; some available methods are—`'linear'` (default), `'pchip'` (called `'cubic'` in early MATLAB® versions), and `'spline'`; a default method that does not need to be specified when the first command configuration is used;

The `'extrap'` string is required for extrapolation by the `'linear'` method; for the `'pchip'` and `'spline'` methods, the word `'extrap'` can be omitted.

For the example presented in Fig. 4.5, the commands for *T* calculations at the time points 1.3 and 7 s (interpolation), and 0.5 and 9.5 s (extrapolation) are:

```
>> t=[1, 3, 5, 9];                          %times, the data for x
>> T=[700, 503, 402, 360];                  % temperatures, the data for y
>> ti=[1.37];                               %1.3 and 7 are the points of interpolation
>> te=[.5 9.5 ];                            %0.5 and 9.5 are the points of extrapolation
>> T_i=interp1(t,T,ti,'spline');            %defines interpolated values
>> T_e=interp1(t,T,te,'spline','extrap');   %defines extrapolated values
>> T_i,T_e                                  % display inter- and extrapolated values
   T_i =
      663.0720   365.0000
   T_e =
      768   360
```

4.3.2 Nonlinear algebraic equation solution

Matrix solution of the linear algebraic equations, i.e., defining unknowns, was discussed in the second chapter (Sections 2.2.2 and 2.3.4.1). Here we describe a command tool that searches iteratively the *x* value in a nonlinear single equation $f(x)=0$. The function for this process is `fzero`, and its general form is:

```
x=fzero('fun', x0)
```

where the `'fun'` is the string with the name of the user-defined function containing the solving equation, or the string where the solving equation can be written; `x0` is the initial or so-called guess value representing a priori given *x*-value, around which the command seeks the true x.

The approximate value of the `x0` can be determined from the *f(x)* graph by checking the *x*-value at which the function is zero.

To demonstrate the command usage, consider the inverse problem for the dielectric constant ε equation of a fluid at the temperature range 0.1–99°C

$$\varepsilon = 87.74 - 0.4T + 9.4 \cdot 10^{-4}T^2 - 1.41 \cdot 10^{-6}T^3$$

where *T* is the temperature in degrees Celsius. If, for example, the required dielectric constant value is 78.3, what is the temperature that meets the required value?

The equation should be rewritten in form $f(x) = 0$:

$$\varepsilon - 87.74 + 0.4x - 9.4 \cdot 10^{-4}x^2 + 1.41 \cdot 10^{-6}x^3 = 0.$$

here the temperature *T* is signed as *x*. The initial temperature value *x*0 must be within the actual *T*-range, for our case—49°C.

The solution command is:

```
>> Temperature=fzero('78.3-87.74 +0.4*x-9.4e-4*x.^2+1.41e-6*x.^3',49)
Temperature =
   25.0154
```

The string with the solved equation cannot include preassigned variables, e.g., it is not possible to define epsilon=78.3 and then write fzero('epsilon-87.74 +0.4*x-9.4e-4*x.^2+ 1.41e-6*x.^3',49).

The following `fzero` form is available for substitution of one or more of preassigned parameters into the solving expression:

```
x=fzero(@(x) fun(x, preassigned_variables1,
preassigned_variables2,... ), x0)
```

where `fun(x, preassigned_variables1, preassigned_variables2,...)` is the solving equation with the searching argument x and preassigned the `preassigned_variables1, preassigned_variables2,...` arguments. The `@(x)`[1] symbols denote that next to these signs the `fun` is a function of x; the fully written `fun` function can be noted e.g., as `f` (see below) and written in a separate line; in this case, the function looks as `x=fzero(f,x0)`.

Using the latter `fzero` form, we can write the above example as a function file with the name `dielconst`, input parameters `epsilon` and `x0`, and output parameter `Temperature`. The full text of the `dielconst` function is:

```
function Temperature=dielconst(epsilon,x0)
% calculates the temperature for given dielectric constant
% to run: >>Temperature=dielconst(78.3,49)
f=@(x) epsilon-87.74 +0.4*x-9.4e-4*x.^2+1.41e-6*x.^3;
Temperature=fzero(f,x0);
```

---

[1] The symbol @ is used in MATLAB® to denote the so-called anonymous functions.

This function should be saved in the file named `dielconst.m` and run in the Command Window with the assigned arguments `epsilon`, and `x0`:

```
>> Temperature= dielconst (78.3,49)
Temperature =
    25.0154
```

Note: the form `f=@(x) fun(x, preassigned_variables1, preassigned_variables2, ...)` can be used for calculating `f` at some `x` values; in this case, after the form `f=@(x)` was written, the `fzero` command calculates the `f` values at the given `x` (see example in the next subsection).

4.3.3 Finding of the extremal points of a function

In materials science/engineering/technology practice, it is necessary to find extremum of the variable (maximum or/and minimum) from the governing equation in order to achieve its best value.

The command destined for these purposes is `fminbnd`. The command search the minimal x value for one-variable function $f(x)$; it has the following simplest forms:

```
x=fminbnd('fun',x1,x2)   and   [x,f_x]=fminbnd('fun',x1,x2)
```

where:

> `fun` is the equation, in which the minimum is searched, written as string between the single quotations (appears in blue); alternatively, the equation can be written after the `@(x)` symbols;
> `x1` and `x2` are the ends of interval where the x-value should be defined;
> `x` and `f_x` are the output parameters containing minimum point values x_{min} and f_{min} (the f-value at minimal x).
> The `fminbnd` command also can be used to find the maximum point of the function by finding the minimum of the function multiplied previously by −1.

To demonstrate this command implementation, find the minimum and maximum of a modified Buckingham equation for the pair-wise interatomic potential:

$$v = Ae^{-\frac{r}{b}} - \frac{C}{r^6}$$

where r is distance between the two atoms; A, b, and C are constants.

Problem: Find the minimal and maximal r and v values of the Buckingham potential; take $A = 6.482 \cdot 10^4$ eV, $b = 0.16$, and $C = 117.3$ eV/A^6; the r range is 0.8 … 4 A for minimal point search and 0.8 … 1.5—for maximal point. Plot the $v(r)$ curve together with defined extremal points; add captions and legend. Write the program in form of the user-defined function named `BuckExtram` with A, b, C, $x1$, $x2$, and $x3$ (searching interval x-boundaries) as input parameters and r_{min}, f_{min}, r_{max}, and f_{max} as output parameters.

The commands that solved this problem are:

```
function[r_min,f_min,r_max,f_max]=BuckExtram(A,b,C,x1,x2,x3)
%finds minimal r and f(r) for Buckingham potential
%to run>>[r_min,f_min,r_max,f_max]=BuckExtram(6.482e4,0.16,117.3,0.8,4,1.5)
close all
f1=@(r)A*exp(-r./b)-C./r.^6                    % function for minimum
[r_min,f_min]=fminbnd(f1,x1,x2);                % minimum search
f2=@(r)(-1)*(A*exp(-r./b)-C./r.^6);             % reverse f1 by (-1)*f1
[r_max,fmax]=fminbnd(f2,x1,x3);                 % maximum via minimum search
f_max=(-1)*fmax;                                % reverse minimal point to maximal
rg=linspace(x1,x2);                             % vector rg for the x-coordinates of the v(r) graph
vg= f1(rg) ;                                    % vector vg for the y-coordinates of the v(r) graph
plot(rg,vg,r_min,f_min,'o',r_max,f_max,'s')     % plot v(r) and the extremal points
grid on
xlabel('Distance r, A'),ylabel('Potential v, eV')
title('Buckingahm potential')
legend('v(r)','minimal point','maximal point')
```

Save this function in the BackExtram file and run it:

```
>> [r_min,f_min,r_max,f_max]=BuckExtram(6.482e4,0.16,117.3,0.8,4,1.5)
r_min =
   1.3670
f_min =
   -5.3515
r_max =
   0.9047
f_max =
   13.0807
```

Resulting plot is:

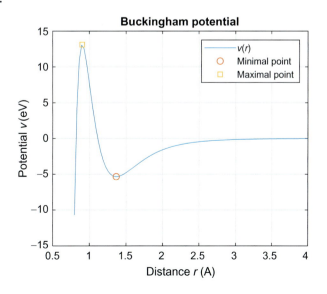

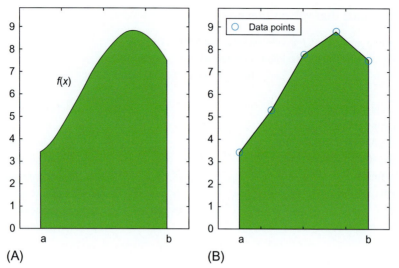

FIG. 4.6 The f(x) function given analytically (A) and by the data points (B); the *shaded area* represents the definite integral.

Note: the `fun` should be written using element-wise operators, e.g., ".*", ".^", and "./".

4.3.4 Integration

The integrals appear in a variety of practical situations and can be solved with the MATLAB® commands, which calculate the integral numerically as the area beneath the function. The function f(x) can be presented by mathematical expression or by tabulated data (Fig. 4.6A and B). For numerical calculations of the integral, the area beneath the f(x) curve is subdivided in small identical geometrical elements, e.g., rectangles or trapezoids. The sum of areas of these elements gives apparently the integral with some uncertainty.

Below, the `quad` and `trapz` functions of integration are described. The first is used when an integrating function is presented as an analytical expression, and the second—when it is presented as data points.

4.3.4.1 The quad function

The curve to be integrated numerically can be approximated by three-point parabolas, and the areas beneath them are calculated and summarized; this method is called Simpson's rule. The function uses an adaptive Simpson's rule of integration and has the form:

```
I=quad('function',a,b,tolerance)
```

where

`'function'` is a string written in single quotations and containing the expression f(x) to be integrated; integrating expression in this place can be written also not as string if the @(x) sign is written before the expression. It is possible also to write here the name

of a function file where the expression should be written; in this case, the @ sign should be written before the name;
a and b are the lower and upper limits of integration;
tolerance—is the desired maximal absolute error; this parameter is optional and can be omitted; the default tolerance is smaller than $1 \cdot 10^{-6}$;
I is the variable to which obtained value of the integral is assigned.

The function *f(x)* should be written using element-wise operators, e.g., ".*", ".^", and "./".

For example, to calculate the integral $I = \int_0^5 \left(1/\left(1 - e^{-\frac{1}{T}}\right)\right) dx$ (exponential functionarises: in thermodynamics of materials e.g., in the theoretical expressions for low-temperature heat capacity of solids, in material kinetics; in intermolecular potentials; in the normal, Weibull and some other distributions; etc.)—the following command should be typed and entered from the Command Window:

>>I=quad('1./(1-exp(-1./x))',0,5) %Note: the element-wise operations are used
I =
  15.3348

the command can also be written as:

>> I=quad(@(x) 1./(1-exp(-1./x)),0,5)
I =
  15.3348

This form, as stated above (see Section 4.3.2), is suitable when the expression has parameters that should be inputted or preassigned.

Another way to use quad is to write in the Editor window and save the integrating function in the function file. For example:

function y=Ch4_Integr(x)
y=1./(1-exp(-1./x));

For integrating the function saved in the Ch4_Integr.m file, the following quad command should be typed and then entered from the Command Window:

```
>> I=quad(@Ch4_Integr,-3,3)
I =
   15.3348
```

4.3.4.2 The trapz function
The form of this function is:

q=trapz(x,y)

where x and y are vectors of the point coordinates and q is variable with obtained integral value; the function uses the trapezoidal method for numerical integration when the function is presented as data points.

For example, the entropy of a substance should be defined at $T = 120\,K$ by the table with heat capacity values measured at temperatures close to absolute zero.

T (K)	1	20	40	60	80	100	120
C_p (J/(mol K))	0.000743	0.462	3.740	8.595	12.85	16.01	18.25

The entropy S is approximately:

$$S = \int_{1}^{120} \frac{C_p(T)}{T} dT$$

where $C_p(T)$ under the integral is given numerically in the table.

The following commands should be typed and entered from the Command Window to find the integral value:

```
>>T=[1 20:20:120];
>>Cp=[.000743 .462 3.74 8.595 12.85 16.01 18.25];
>>S= trapz(T,Cp./T)
S =
   13.1278
```

4.3.5 Derivative calculation

The derivative $\frac{dy}{dt}$ of the function $y(t)$ at some given t point can be approximately represented as the ratio of infinitesimal changes of function Δy to its argument Δt. In geometrical representation (Fig. 4.7), it is the slope $\alpha \approx \frac{\Delta y}{\Delta t}$ of the tangent line to the curve at the i-th point.

Following these definitions, the derivative at a point i can be presented as:

$$\frac{dy}{dt} = \lim_{\Delta t \to 0} \frac{\Delta y}{\Delta t} \approx \frac{y_i - y_{i-1}}{t_i - t_{i-1}}$$

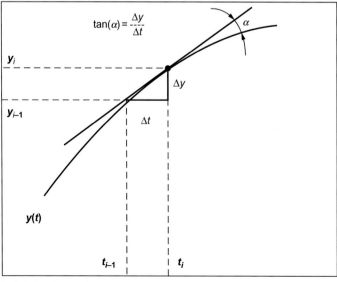

FIG. 4.7 Derivative dy/dt in geometrical representation.

In this interpretation, the derivative can be calculated at each $(i-1)$-point as the ratio of y- and x-differences between neighboring points i and $i-1$.

To calculate the differences Δy or Δt, the `diff` command can be used; it has the following forms:

```
dy=diff(y)     or    dy_n=diff(y,n)
```

where `y` is the inputting vector with y-values that numbered corresponding to points $i=1, 2, \ldots$; the `n` argument indicates how many times the `diff` should be applied and the order of the particular difference. E.g., if `n=2` the `diff` should be applied twice; the `dy` and `dy_n` denotes the output parameters and are vectors of the determined values of the 1st and n-th order differences.

Note: the `dy` vector is 1 element shorter than vector `y`, and correspondingly the `dy_n` vector is n elements shorter than `y`; e.g., if `y` has 10 elements, `dy` has 9 elements and, say, the second order derivative vector `dy_2` has only 8 elements.

Analogously to integration, the derivative can be calculated for function given analytically or as a table. In the latter case, the number of points that can be used for calculations is limited by the table, and the derivative value may be less accurate than in the case of an analytically given function, when you can specify a greater number of points to ensure greater accuracy. For example, coefficient of linear expansion L of a metal due to temperature is $\alpha_L = \frac{1}{L}\frac{dL}{dT}$. The length-temperature data for the Al-rod of 2.4 m length at $T=20°C$ is shown in the table.

T (°C)	50	100	150	200	250	300
L (m)	2.401728	2.404608	2.407488	2.410368	2.413248	2.416128

These data have to be used to determine the thermal expansion coefficient at 50 ... 250 T. The commands to be applied in the Command Window are:

```
>> T=[50 :50 :300];
L=[2.4018 2.40461 2.40749 2.41037 2.41325 2.4161];
dT=diff(T);
dL=diff(L);                                        % T and L differences
dLdT=dL./dT;                        % calculates derivative, uses element-wise division
alpha=dLdT./L(2:end);                   % coefficient of the thermal expansion
disp('T     L     Alpha')
fprintf('%3i  %7.4f%11.8f\n',[T(2:end); L(2:end); alpha])

    T    L       Alpha
    100  2.4046  0.00002337
    150  2.4075  0.00002393
    200  2.4104  0.00002390
    250  2.4133  0.00002387
    300  2.4161  0.00002359
```

Note, that if the step h of argument of the differentiating function is the constant value, it can be used instead of diff(T).

In the example above, the function was given by the table; nevertheless, the derivatives of a function given by an equation can be determined in the same way as from the tabulated data; the step in argument in such a case can be smaller and leads to more exact derivative values. For example, assuming that the data in the above table are very accurately described by the equation $L = 2.3988 + 5.76 \cdot 10^{-5} * T$, the coefficient of thermal expansion can be determined with the following commands:

```
>>h=25;                          % L differences, and also step
T=50:h:300;
L=2.3988+5.76e-5*T;%
dL=diff(L);                      % L differences
dLdT=dL./h;                      % calculates derivative, uses element-wise division
alpha=dLdT./L(2:end);            % coefficient of the thermal expansion
disp('T    L    Alpha')
fprintf('%3i %7.4f%11.8f\n',[T(2:end); L(2:end); alpha])
  T     L       Alpha
  75  2.4031  0.00002397
 100  2.4046  0.00002395
 125  2.4060  0.00002394
 150  2.4074  0.00002393
 175  2.4089  0.00002391
 200  2.4103  0.00002390
 225  2.4118  0.00002388
 250  2.4132  0.00002387
 275  2.4146  0.00002385
 300  2.4161  0.00002384
```

The implementation of the analytical expression $L(T)$ made it possible to obtain a greater number of the exact α_L values than using the table, when there are not sufficient points to correct derivative calculation.

Frequently, a problem is formulated so that the derivative should be determined for one specific point. For this problem, in case of analytically given function, the minimal range of argument should be assigned. For example, if the required derivative value is needed at argument point $t=1$, the minimal two points for the numerical differentiation is t and $t-h$, where h must be the specified argument step.

4.3.6 Supplementary commands for interpolation, equation solution, integration, and differentiation

In addition to commands discussed above, Table 4.1 shows another function that can be used for inter-/extrapolation, solution the nonlinear equations, integration, and derivative calculations.

Table 4.1 Some commands for inter- extrapolation, solution the nonlinear equations, integration, and derivative calculations.[a]

MATLAB® command	Description	Example of usage (input and output)
yi=spline(x,y,xi)	interpolates with the cubic splines. x and y are the vectors with the inputting argument and function data; xi is the x-point/s of interpolation; yi— outputted interpolated data.	>>x=[0 1 2.5 3.6 5 7]; >>y=sin(x);xi=2.7; >>yi=spline(x,y,xi) yi = 0.4297
yi=pchip(x,y,xi)	Piecewise Cubic Hermite Interpolating Polynomial (PCHIP). x, y, xi, and yi—the same as at the spline command	>>x=[0 1 2.5 3.6 5 7]; >>y=sin(x);xi=2.7; >>yi=pchip(x,y,xi) yi = 0.4854
x=roots(p)	returns the roots x of the polynomial represented by the column vector p containing the polynomial coefficients. The polynomial has form: $p_1 x^n + \ldots + p_n x + p_{n+1} = 0$	% find roots of the %polynomial % 2.6x^2-1.5x -3.9=0 >> p = [2.6 -1.5 -3.9]; >> r = roots(p) r = 1.5467 -0.9698
x=fminsearch(fun,x0)	finds minimum x of multivariable nonconstrained function fun; the calculations begin from the x0 value.	>>f=@(x)100*(x(2) - x(1)^2)^2 + (1 - x(1))^2; >>x0=[-1.2,1]; >>x = fminsearch(f,x0) x = 1.0000 1.0000
I = integral(fun,a,b)	integrates expression written in the fun; a and b are the lower and upper limits of integration respectively; returns numerical value of the integral in the output variable I.	>>A=3;a=-3;b=3; >>f=@(x)A*exp(-x.^2); >>I=integral(f,a,b) I = 5.3172
I=quadl(fun,a,b)	returns numerical value I of the integral; fun is the integrating expression written as string or name of subfunction containing the expression; a and b are the lower and upper limits of integration respectively.	>> I=quadl('3*exp(-x.^2)', -3,3) I = 5.3172
I=quadgk(fun,a,b)	Returns integral of the fun function. All notations are the same as in the integral, but one of the limits can be infinite (noted inf).	>> f = @(x) exp(-x.^2); >> Q = quadgk(f,0,Inf) Q = 0.8862
I=quad2d(fun,a,b,c,d)	returns numerical value I of the double integral; fun is the integrating expression written as string or as the subfunction containing the expression; the integration limits a and b for the first variable (i.e., x) and c and d - for the second (i.e., y).	>>f = @(x,y) y.*sin(x); >>I = quad2d(f,pi,2*pi,0,pi) I = -9.8696

Continued

Table 4.1 Some commands for inter- extrapolation, solution the nonlinear equations, integration, and derivative calculations[a]—cont'd

MATLAB® command	Description	Example of usage (input and output)
dcoeff=polyder (p)	calculates coefficients (returns in dcoeff) of the polynomial that is derivative of the polynomial having the p-coefficients; p has the same sense as in the root command.	% derivatives of the %polynomial % 2.6x^2-1.5x -3.9=0 >>p=[2.6 -1.5 -3.9]; dcoeff=polyder(p) dcoeff= 5.2000 -1.5000 % coefficients in the %derivative polynomial % 5.2x-1.5=0

[a] Commands are given in their simplest form.

4.4 Live Editor

From the R2016a MATLAB® version, the Live Editor was announced—a tool allowing to produce interactive script programs that include commands, explanatory text, and images together with numerical and graphical results of program executions. Beginning from the R2018a version, the Live Editor possibilities were extended, and now the user-defined functions can be created in the Live Editor. Here, we bring a brief description of the Live Editor and creation of the live script file containing a MSE-oriented example.

4.4.1 Launching the Live Editor

To open the Live Editor—click the New Live Script button located in the File section of the MATLAB® Desktop menu. The Live Editor window appears—Fig. 4.8.

The toolstrip has three tabs—Live Editor, Insert, and View. The Live Editor tab contains the sections with buttons that are used to edit the scripts. The Insert tab contains functions to insert texts, images, equations, section breaks, controls, and others to insert items into the script. The View tab includes functions to the live script output. Below, we describe those of the tabs buttons that are necessary to use for creating actual live script.

The Live Editor window contains the vertical dividing line (default) that separates the window into two parts: the left—for entering commands, explanatory texts, equations, and images, and the right—to show output numbers and graphs. To show outputs inline, not on the right, the "Show outputs inline" button should be clicked.

Chapter 4 • Writing programs for technical computing 145

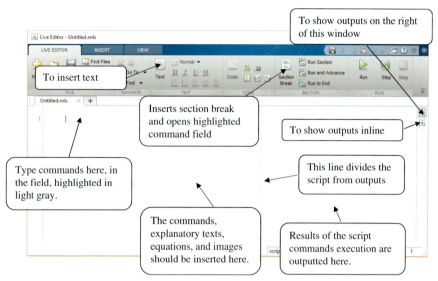

FIG. 4.8 The Live Editor windows.

4.4.2 Creating the life script with the Live Editor

We demonstrate the work with the Live Editor by example. Build the script that calculates and plots the stresses fields near the crack-tip of a linear elastic material for the mode 1 loading. The appropriate equations are:

$$\sigma_{xx} = \frac{K}{\sqrt{2\pi r}} \cos\frac{\theta}{2}\left[1 - \sin\frac{\theta}{2}\sin\frac{3\theta}{2}\right]$$

$$\sigma_{yy} = \frac{K}{\sqrt{2\pi r}} \cos\frac{\theta}{2}\left[1 + \sin\frac{\theta}{2}\sin\frac{3\theta}{2}\right]$$

$$\tau_{xy} = \frac{K}{\sqrt{2\pi r}} \cos\frac{\theta}{2}\sin\frac{\theta}{2}\cos\frac{3\theta}{2}$$

where K is the stress intensity factor, r and θ denote respectively the distance from the crack edge to a point in the stress field and angle between r direction and positive direction of the x-axis.

Problem: Create the live script that contains the image illustrating the crack, three above equations, commands for calculating the stresses σ_{xx}, σ_{yy}, τ_{xy} and three outputted three-dimensional plots for each of the stresses. Take $K = 540\,\text{MPa m}^{1/2}$, $r = 0.0005$, 0.0006, ..., 0.005 m, $\theta = 0$, $\pi/20$, ..., $\pi/2$.

After opening the Live Editor, the following steps should be performed to solve the problem.

Step 1. Inserting the title/text.

For inserting the following title "Stress fields near a crack tip. Linear elastic isotropic material, mode 1," click the Text button of the Live Editor tab. In the white space appearing above the light gray command field, type the above caption.

Step 2. Inserting the image that illustrates the crack and stresses.

If the necessary image was previously created and saved as jpg-, bmp-, gif-, png-, or wbmp- file, then click the Image button of the "Insert" tab. In the "Load image" window, the containing image file should be selected and the Open button clicked. As result, the selected image appears in the text editor field.

Step 3. Inserting the equations.

To insert the solving equations as text, select the Equation button of the "Insert" tab.

The equations should be typed into the "Enter your equation" field that appears under the inserted image. In the toolstrip, the additional tab named Equation appears; the tab allows inputting the necessary symbols and mathematical structures to type the equations.

Step 4. Entering commands for stress fields calculations.

Before entering these commands, insert the explanatory text—"Calculations by the above equations" (in the same way as described in Step 1). Then, the following commands should be typed within the highlighted command field.

```
K=540;r=0.0005:0.0001:0.005;Tet=0:pi/20:pi/2;
[x,y]=meshgrid(r,Tet);
sxx=K./sqrt(2*pi*x).*cos(y./2).*(1-sin(y./2).*sin(3*y./2));
syy=K./sqrt(2*pi*x).*cos(y./2).*(1+sin(y./2).*sin(3*y./2));
sxy=K./sqrt(2*pi*x).*cos(y./2).*sin(y./2).*cos(3*y./2);
```

Step 5. Enter in separate sections the text captions and commands for each of three stress plots.

Click the Section Break button, the horizontal separating line appears together with highlighted field for entering commands. To enter the "Plotting the σ_{xx}" text before commands, the same procedures should be followed as described in Step 1. For generating and formatting plot, the following commands should be entered into the command field of the first section:

```
surf(x,y,sxx)
xlabel('r'), ylabel('\Theta'), zlabel('\sigma_x_x')
```

Click the Section Break button and enter text "Plotting σ_{yy}" and then the commands as follows:

surf(x,y,syy)
xlabel('r'),ylabel('\Theta'),zlabel('\sigma_y_y')

Analogously, create the third section, enter text "Plotting τ_{xy}" and commands:

surf(x,y,sxy)
xlabel('r'),ylabel('\Theta'),zlabel('\tau_x_y')

Step 6. Running the live script.

To run the live script, click the Run button located in the Live Editor tab of the toolstrip. After some time slightly greater than when running the simple scrip, the three plots are outputted on the right side of the Live Editor window. All live script programs with outputted plots are represented in Fig. 4.9.

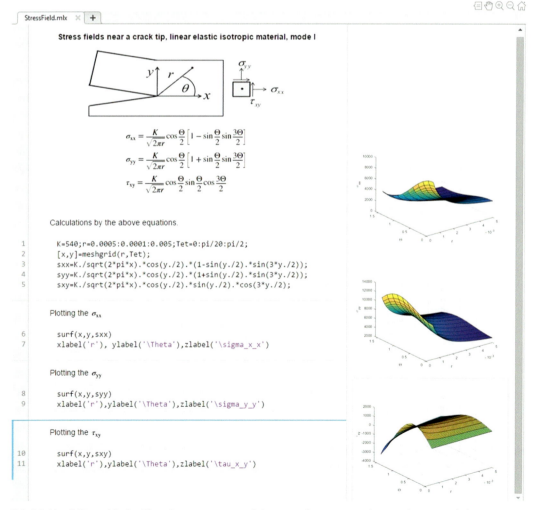

FIG. 4.9 Live Editor with the life script program containing texts, images, equations, and outputted plots.

Note, to run the section of the script, click the blue bar to the left of the section; or optionally, click the Run Section button located in the "Section" section of the Live Editor tab.

Step 7. Saving created live script.

The script can be saved in a file by clicking the Save button on the toolstrip and giving a file name in the opened "Select file for Save as" window; the saved script file gets the *mlx* extension. So the life script file shown in Fig. 4.9 is named StressField.mlx.

For publishing, the script can be rearranged in the pdf-file by selecting the "Export to pdf …" line in the popup menu of the Save button.

4.4.3 Supplementary information for using the Live Editor

4.4.3.1 Creating Live Function

The steps described above produce a live script. To create a live function, select the Live Function line in the popup menu of the New button located in the Home tab of the MATLAB® Desktop. The Live Editor opens with the untitled function (default)—Fig. 4.10.

Note, the functions in the Live Editor has the `end` as its last line; this form is accepted for so-called nested functions that are not studied in this book. The final `end` can be omitted in case of user-defined function generated by the rules described in this chapter (Section 4.2).

To run the live file function, enter its name with input parameter values in the Command window; numerical and graphical results appear in the Command Window and Figure window, respectively.

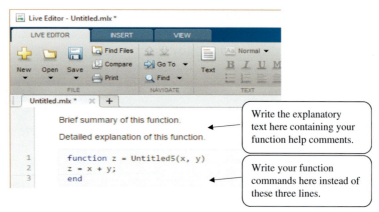

FIG. 4.10 Live Editor opened for creating the Live Function.

4.4.3.2 Opening the existing script as a live script and user-defined function as a live function

The existing m-file containing script can be opened as a live script (.mlx). For this, when the file is opened in the Editor, click on the Save button of the Editor tab; select Save As line in the popup menu, and in the "Save as type" field of the opened window, select the "to MATLAB® Live Code Files (*.mlx)" option and click Save. Opening/saving a script as a live script creates a copy of the original script. The comments of the script are converted into the texts of the generated live script.

The existing m-file containing function can be opened as a live function (.mlx) file in the same way as for a script file with only one difference—selecting the "to MATLAB® Live Code Files (*.mlx)" option instead of "to MATLAB® Live Code Files (*.mlx)" (see above).

4.4.3.3 Text formatting options

To format existing text, the following Text section option of the Live Editor should be used:

- Text Style: Normal, Heading 1, or 2, or 3, Title;
- Alignment: Left, Center, Right;
- Lists: Numbered list, Bulleted list;
- Standard Formatting: Bold, Italic, Underline, Monospace;
- Change Case (select the text and right click): uppercase, lowercase.

4.4.3.4 About the interactive controls

For interactive control variables, the Live Editor has the sliders and dropdowns that can be added to live script or function. To insert slider—click the Control button of the Insert tab and select Numeric Slider line; then specify a minimum, maximum, and step values of the variable. The same steps should be performed for inserting the drop-down; in this case, the Drop-Down line should be selected, and the desirable values should be inputted.

For example, in the discussed above live script, the liner for stress intensity factor *K* looks like K= 460 . The value to the left is the current value of the *K*. In case the drop-down control was selected, the command for *K* looks like—

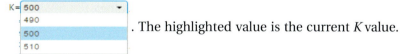
. The highlighted value is the current *K* value.

Note: Live Editor is a new and more suitable tool for interactive programming than the MATLAB® Editor, but life scripts and programs work much slower than regular scripts and functions, and has limitations in the use of certain classes of variables. All this makes it difficult to use the programs created with the Live Editor for problems with a large volume of data as well as in programs with a large number of commands. Thus, the user should choose the editor orienting to his specific problem. Therefore, in the book the regular MATLAB® editor is used.

4.5 Application examples

4.5.1 Convertor from Brinell to Vickers and Rockwell hardness

In tests of the material hardness *H*, its values are obtained in different units—Brinell, Vickers, Rockwell B, Rockwell C, etc. To coincide results, the data should be converted to the same hardness scale. There is no simple expression for conversions; thus, in this example, the approximate expressions were taken. The expressions convert from Brinell, *HB*, to Rockwell C, *HRC*, and Vickers, *HRB*, hardness:

$$HV = \frac{HB}{0.95}$$

and

$$HRB = 134 - \frac{6700}{HB}$$

Problem: Compose a script program that converts from Brinell to both Vickers and Rockwell B hardness by the expressions above. The program should request the input of the Brinell hardness value and display the calculated *HB* and *HV* hardness in line with *HB* hardness with one decimal digit. Save program in the `ApExample1.m` file

The program that solved this problem is:

```
% HB to HV and HRB convertor
% HB can be single number or vector, e.g. [502 473 150]
HB=input('Enter hardness, in the Brinel's units ');
HV=HB/0.95;                              % HB to HV
HRB=134-6700./HB;                        % NB to HRB
Results=[HB;HV;HRB];                     % three-row matrix
fprintf('\n %5.1f HB  is %5.1f in HV and %5.1f in HRB\n',Results)
```

When the script file name—`ApExample1`—is entered into the Command Window, the program displays the prompt to input the data. With the hardness entered as vector (in square brackets with spaces between the numbers) and after the Enter key pressed, the HB, `HV`, and `HRB` are calculated and displayed with one digital number after the decimal point.

The following shows the Command window when the file is running:

```
>> ApExample1
Enter hardness, in the Brinell's units [500 478 192]
500.0 HB is 526.3 in HRB and 120.6 in HV
478.0 HB is 503.2 in HRB and 120.0 in HV
192.0 HB is 202.1 in HRB and 99.1 in HV
```

4.5.2 Dynamic viscosity: Reyn to pascal-second and vice versa convertor

The British units of dynamical viscosity is **reyn** (lb s/in.^{2}). To convert it to the International System of units (SI)—pascal-second (Pa s) and vice versa—form SI to British—the following relations are used: 1 Pa s = 1/6890 reyn and 1 reyn = 6890 Pa s,

Problem: Create a script program that converts the dynamic viscosity values from British units to the pascal-second and vice versa. The program requests what conversion you desire to execute: inputting the number 1 means reyn to Pa s conversion, any other number—the Pa s to reyn. After this, the program requests the viscosity value in appropriate units and displays converted values in Pa s or in reyn. Save program in m-file named `ApExample2`.

To solve this problem, carry out the following steps:

- the program displays a two-line message—Enter 1 for reyn to Pa s or other number for Pa s to reyn—after this, the number 1 should be inputted for the reyn to Pa s conversion, or number 2 or some other for the Pa s to reyn conversion;
- in accordance to selected conversion, the program displays message "Enter viscosity in the reyn units" or "Enter viscosity in the Pa s units"; after this, the single value or number converting values in brackets should be entered;
- the program converts entered value to the desired units and displays the inputted and converted values using the fprintf command, with the single f sign in its number formatting parts.

The program solving the problem is written below.

```
% reyn to Pa.sec and vice versa convertor
% enter the 1 number for rey to Pa.s, and 2 or other number for Pa.s to reyn
%         enter single number or vector e.g [84 540 3200], in Pa.s
%     enter single number or vector e.g [.58e6 3.72e6 2.2e+07], in Pa.s
cond=input('Enter 1 for reyn to Pa.s \nor other number for Pa.s to reyn ');
if cond==1                                        % for reyn to Pa.s
    reyn=input('Enter viscosity in the reyn units ');
    Pa_s=reyn/6890;
    Results=[reyn;Pa_s];                          % two-row matrix
    fprintf('Viscosity is %f in reyn and %f in Pa.s\n',Results)
else                                              % for Pa.s to reyn
    Pa_s=input('Enter viscosity in the Pa.s units ');
    reyn=6890*Pa_s;
    Results=[Pa_s;reyn];                          % two-row matrix
    fprintf('Viscosity is %f in Pa.s and %f in reyn\n',Results)
end
```

The running command, inputted numbers, and displaying results for the reyn to Pa s conversion:

```
>> ApExample2
Enter 1 for reyn to Pa.s
or other number for Pa.s to reyn 1
Enter viscosity in the reyn units [.54 0.986 1.49]
Viscosity is 0.540000 in reyn and 0.000078 in Pa.s
Viscosity is 0.986000 in reyn and 0.000143 in Pa.s
```

```
Viscosity is 1.490000 in reyn and 0.000216 in Pa.s
```
And for Pa·s to reyn conversion
```
>> ApExample2
Enter 1 for reyn to Pa.s
or other number for Pa.s to reyn 2
Enter viscosity in the Pa.s units [78e-6 143e-6 216e-6]
Viscosity is 0.000078 in Pa.s and 0.537420 in reyn
Viscosity is 0.000143 in Pa.s and 0.985270 in reyn
Viscosity is 0.000216 in Pa.s and 1.488240 in reyn
```

4.5.3 Thermal conductivity data interpolation

The data obtained in a material properties experiment are given as table of the measured points. Frequently, it is need to estimate values between the points or out of measured range. Take, as an example, the measured data on the thermal conductivity-temperature of a metal, presented in the table below.

T (K)	173	273	373	573	973
λ (W/(m K))	420	405	395	381	354

Problem: write user-defined function ApExample3 containing the tabular data and commands for its interpolation; assign the interpolating temperature as input parameter and calculated thermal conductivities as output parameter. Try the linear, pchip (former "cubic"), and spline methods for points between the data—200, 300, 500, and 800 K. Plot the original and interpolated points for each method placing on the same page (Figure window). Save the program in the function file named ApExample3.

The function is arranged as follows:

- the first line defines function name, ApExample3, and input and output variables— T_interp and Labda_interp respectively;
- two following lines present the help part of the function and contain its short purpose and command for the function run in the Command Window;
- in the next lines, the temperatures and thermal conductivities data is assigned to variables T and Lambda respectively; then the interp1 commands for each of three interpolating methods assign the interpolating values to the three-rows output matrix Labda_interp. in such a way that each row contains all values interpolated by the same method;
- further lines represent commands for generating and formatting three plots in the same Figure window; the tabular data signed by circles and connected with the solid line and interpolating values signed by the purple x (magenta color); each plot is formatted with the axis labels, grid, a caption, and a legend.

The commands that solved the problem are:

```
function Lambda_interp=ApExample3(T_interp)
% Thermal conductivity data interpolation
% to run >>Lambda_interp=ApExample3([173:100:973])
T=[173 273 373 573 973];                                         %Temperature vector
Lambda=[420 405 395 381 354];                                    %Termal conductivity vector
Lambda_interp(1,:)=interp1(T,Lambda,T_interp,'linear');          %1st Lambda_interp row
Lambda_interp(2,:)=interp1(T,Lambda,T_interp,'pchip');           %2nd Lambda_interp row
Lambda_interp(3,:)=interp1(T,Lambda,T_interp,'spline');          %3rd Lambda_interp row
subplot(1,3,1)
plot(T,Lambda,'o-',T_interp,Lambda_interp(1,:),'xm'),grid on     %1st plot
xlabel('Temperature, K')
ylabel('Thermal conductivity,W/(m K)')
title('Linear interpolation')
legend('Data','Interp.','location','southwest')                  % left bottom location
subplot(1,3,2)
plot(T,Lambda,'o-',T_interp,Lambda_interp(2,:),'xm'),grid on     %2ndplot
xlabel('Temperature, K')
ylabel('Thermal conductivity,W/(m K)')
title('PCHIP interpolation')
legend('Data','Interp.','location','southwest')                  % left bottom location
subplot(1,3,3)
plot(T,Lambda,'o-',T_interp,Lambda_interp(3,:),'xm'),grid on     %3rd plot
xlabel('Temperature, K')
ylabel('Thermal conductivity,W/(m K)')
title('Spline interpolation')
legend('Data','Interp.','location','southwest')                  % left bottom location
```

To run the program, the command written in the function help should be typed in the Command Window. The running command, outputted values of the thermal conductivity, and generated plots are presented below.

```
>> Lambda_interp=ApExample3([200 300 500 800])
Lambda_interp =
 415.9500 402.3000 386.1100 365.6775
 415.4306 401.9271 385.8398 365.5266
 415.2857 401.9332 385.5513 367.5025
```

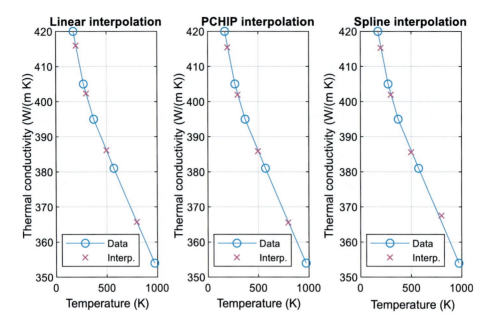

4.5.4 Density of a solid at given pressure and temperature

A relation between density, pressure, and temperature of solids may be described with the Mie-Grüneisen equation of state:

$$p = \frac{\rho_0 C_0^2 \chi \left(1 - \frac{\Gamma_0}{2}\chi\right)}{(1 - s\chi)^2} + \Gamma_0 \rho_0 c_v (T - T_0)$$

$$\chi = 1 - \frac{\rho_0}{\rho}$$

where p is pressure in Pa (or J/m^3), ρ and ρ_0 are current and initial density respectively in kg/m^3, c_v—heat capacity in J/(kg K), T and T_0—current and initial temperature respectively in K, C_0 is bulk speed of sound in m/s, Γ_0 and s—solid-specific constants.

Problem: write the user-defined function that obtains solid density at $p = 1.013 \cdot 10^5$ Pa, $T = 600$ K, $\rho_0 = 8.96 \cdot 10^3$, $c_v = 390$, $T_0 = 298$, $C_0 = 3.933 \cdot 10^3$, $\Gamma_0 = 1.99$, and $s = 1.5$ (cooper). Set desired temperature as the function input and the density found as the output. Use the `fzero` function with `x0` equal to $8.9 \cdot 10^3$ for obtaining the density value from the Mie-Grüneisen equation.

For using `fzero`, the variable ρ is renamed as x, and the expressions above should be rewritten in form $f(x) = 0$:

$$\frac{\rho_0 C_0^2 \left(1 - \frac{\rho_0}{x}\right)\left[1 - \frac{\Gamma_0}{2}\left(1 - \frac{\rho_0}{x}\right)\right]}{\left[1 - s\left(1 - \frac{\rho_0}{x}\right)\right]^2} + \Gamma_0 \rho_0 c_v (T - T_0) - p = 0$$

When run, the function should also graphically display the function $f(x)$ with the solution point.

The solution of this problem, written as user-defined function, is:

```
functiondensity=ApExample4(x0)
% calculates density of a solid
% to run >> ro=ApExample4(8.9e3)
close all
p=1.013e5;T=600;T0=298;              % assigns values to the p, T, and T0
ro0=8.96e3;cv=390;C0=3933;           % assigns values to the ro0, cv, and C0
G0=1.99;s=1.5;                       % assigns values to the G0, and s
f=@(x)ro0*C0^2*(1-ro0./x).*(1-G0/2*(1-ro0./x))./(1-s*(1-ro0./x)).^2+...
    G0*ro0*cv*(T-T0)-p;                              % solving equation
density=fzero(f,x0);                 % solves equation with respect to x
xg=linspace(8e3,9e3);
plot(xg,f(xg),ro,f(ro),'o')          % generates plot f(x) with defined points
grid on
xlabel('Density, kg/m^3')
ylabel('f(\ro)')
legend('f(ro)','Solution point','location','southeast')
```

This function is written in the Editor Window and saved as the ApExample4.m function file. The `fzero` command uses the Mie-Grüneisen equation in form @(x) (see Section 4.3.2) that allows substituting the preassigned parameters in the function.

The running command, outputted values of the solid density, and generated plots are:

```
>> density=ApExample4(8.9e3)
density =
 8.8221e+03
```

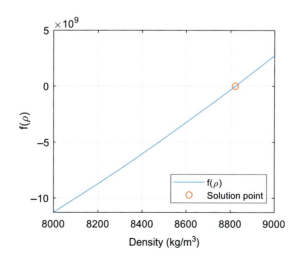

4.5.5 Maximal radiance of a surface by the Planck's law

The radiance of a surface is characterized by the spectral radiance B, kW/m^3 as function of wavelength λ, mm and temperature T, K. In accordance to the Planck's approach, it can be described with the following equation:

$$B = \frac{2 \cdot 10^{27} \pi c^2 h}{\lambda^5} \frac{1}{e^{\frac{1 \cdot 10^6 hc}{\lambda kT}} - 1}$$

where c is the speed of light, equal to $3 \cdot 10^8$ m/s; k is Boltzmann's constant, equal to $1.38 \cdot 10^{-23}$ J/K.

Problem: Create the user-defined function that determines maximum $B(\lambda)$ at $T = 3500$ K with λ in the range 0.2–3 mm. Set desired T as input and defined B as output parameters of the function. Plot graph $B(\lambda)$ and mark the defined maximal point.

The steps for the problem solution are:

- define the function, ApExample5, with input- (T, L1, L2—temperature, and lower and upper wavelengths(c) trapz respectively) and output- (Lmax and Bmax—the wavelength and spectral radiance at the maximum) variables;
- include the help part of the function explaining its purpose and containing line with command for the function run in the Command Window;
- assign the constant values of the solving equation;
- write the Planck's equation after the @(x) for further using in the fminbnd command; as this function searches the minimum, thus the equation should be multiplied by −1; the x denotes the wavelength now;
- write the fminbnd function in the second form described in Section 4.3.3; defined minimal B value should be reverted to represent the maximum;
- further lines should represent commands for generating $B(\lambda)$ plot with the axis labels, grid, a caption, and a legend.
- Add to curve a label showing the temperature value; format the text for this as ['T=' num2str(T) 'K'] vector.

The commands that solved the problem are:

```
function [Lmax,Bmax]=ApExample5(T,L1,L2)
% maximal surface radiance-Planckws law
% [Lmax,Bmax]=ApExample5(3500,0.2,6)
c=3e8;
k=1.38e-23;
h=6.63e-34;
B=@(x)(-1)*(2*pi*c.^2*h./(1e-27*x.^5.)).*1./(exp(1e6*h*c./(x*k*T))-1);
[Lmax,Bm]=fminbnd(B,L1,L2);
Bmax=(-1)*Bm;
Lg=linspace(L1,L2);
Bg=(-1)*B(Lg);
plot(Lg,Bg,Lmax,Bmax,'o')
grid on
xlabel('Wavelength, mm')
ylabel('Spectral radiance, kW/m^3')
legend('B(\lambda)','Maximum')
text(1.8,max(Bg)/2,['T='num2str(T) 'K'])
```

The running command, outputted values of the solid density, and generated plots are:

```
>> [Lmax,Bmax]=ApExample5(3500,0.2,6)
Lmax =
  0.8294
Bmax =
  6.7122e+09
```

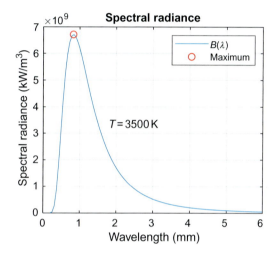

4.5.6 Isochoric thermal pressure coefficient of a substance in the solid state

The isochoric thermal pressure coefficient β_v (denotes the $\left(\frac{\partial P}{\partial T}\right)_v$ derivative) in MPa/K can be defined for a solid at temperatures close to absolute zero through the isochoric heat capacity C_v, J/(mol K):

$$\beta_v = \int_0^{T_u} \frac{1}{T} \left(\frac{\partial C_v}{\partial v}\right)_T dT$$

where T is the temperature in K, v is the volume in m³/mol, and T_u is the upper limit of integration that can vary but should not be equal to the lower limit of integration; the latter is zero in the current case. Relation for $\left(\frac{\partial C_v}{\partial v}\right)_T$ as function of volume and temperature should be known for calculating β_v with the integral above.

Problem: Write the user-defined function that determines β_v when:

$$\left(\frac{\partial C_v}{\partial v}\right)_T = 9.6533 \cdot 10^{-8} e^{0.35 \cdot 10^6 v} T^{3.26}$$

Calculate β_v with the quad function at the following upper integrating limits $T_u = 1, 3, ..., 15$ K for every of the three isochores $22 \cdot 10^{-6}$, $19 \cdot 10^{-6}$, and $18 \cdot 10^{-6}$ m³/mol. Assign the upper temperature limits and step between its as input and resulted thermal pressure coefficient as output value. Generate graph showing the $\beta_v (T)$ isochoric.

The steps to solve the problem are:

- define the function as ApExample6 with the input- (Tu1, Tu2, Ts—the first, latter, and step of upper temperatures) and output- (beta_v—thermal pressure coefficient) variables;
- include the help part explaining the function purpose and containing line with command for the function run in the Command Window;
- assign vector with the upper limits of integration and vector with the isochore values;
- write two the for .. end loops, the external for isochores and the internal for the upper-limit temperatures; in the latter loop, write the isochoric thermal pressure coefficient equation in form @(x) equation and the quad function for integration; the beta_v output parameter of the quad should be indexed accordingly to the Tu and V values for saving all the calculated beta_v;
- put commands for generating β_v plot with the axis labels, grid lines, a caption, and a legend.

```
function beta_v=ApExample6(Ts,Tf,dT)
% calculates the isochoric thermal pressure coefficient by integration
% to run: >>beta_v=ApExample6(1,15,2)
T=Ts:dT:Tf;                                      % assigns temperatures
v=[18e-6 19e-6 22e-6];                           % assigns volumes
for j=1:length(v)                                % volume loop, index j
    for i=1:length(T)                            % temperature loop, index i
        f=@(x)9.5533e-8*exp(0.35e6*v(j))*x.^3.26./x;   % for integration
        beta_v(i,j)=quad(f,Ts,T(i));             % Simpson's integration
    end
end
plot(T,beta_v)
grid on
xlabel('Temperature, K')
ylabel('Coefficient \beta_v,Pa/K')
title('Isochoric Thermal Pressure Coefficient')
legend('v=18 cm^3/mol','v=19 cm^3/mol','v=22 cm^3/mol',...
    'location','best')                           % the best legend location
```

The running command, outputted values of the isochoric thermal pressure coefficient, and generated plot are:

```
>> beta_v=ApExample6(1,15,2)
beta_v =
     0        0        0
  0.0006  0.0008  0.0023
```

```
0.0030 0.0043 0.0122
0.0091 0.0129 0.0368
0.0206 0.0292 0.0835
0.0396 0.0562 0.1606
0.0683 0.0969 0.2769
0.1089 0.1545 0.4416
```

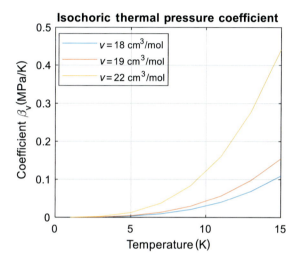

4.5.7 Isothermal compressibility for a gaseous substance

The isothermal compressibility coefficient in atm^{-1} for gaseous substances is determined as:

$$\beta_T = -\frac{1}{v}\left(\frac{\partial v}{\partial p}\right)_T$$

where v is the volume, L/mol, p is the pressure, atm, T is the temperature, K. The derivative in this coefficient can be calculated numerically (see Section 4.3.5) for van der Waals equation of state

$$p = \frac{nRT}{v-nb} - \frac{n^2 a}{v^2}$$

where n is number of moles, a and b are the gas-specified constants in L^2 atm/(mol^2) and L/mol respectively, R is the gas constant equal to 0.08206 in L atm/(mol K).

Problem: Write user-defined function that calculates β_T and having the input parameters only: a, b, n, and T. Take the $a = 3.5924$, $b = 0.04271$, $n = 2$, $T = 293.15$, and v equal to 5, 9, 13, ..., 39. The function should display three columns of the v, p, and β_T values and generate plot $\beta_T (p)$ with the axis labels, grid lines, and a caption.

The steps to solve the problem are:

- define the function as `ApExample7` having the input variables a, b, n, and T (denoted above) and no output variables;
- include the help part explaining the function purpose and containing line with command for the function run in the Command Window;
- assign the gas constant value and vector of the volume values;
- calculate the pressures by the van der Waals equation of state at given constant T and v-values;
- calculate derivative $(dv/dp)_T$ as the ratio `diff(v)./diff(p)` and then calculate the β_T by the equation above;
- display three-column table with the given v, and calculated p, and β_T values;
- put commands for generating $\beta_T(p)$ plot with the axis labels, grid lines, and a caption. Note: to align the length of the p and β_T vectors, the p-vector should begin from its second value.

The commands solving the problem are:

```
function ApExample7(a,b,n,T)
% calculates the isothermal compressibility by differentiation van der Waals
equation of state
% to run: >>ApExample7(3.5924,0.04271,2,293.15)
close all
R=0.08206;
v=5:4:37;
p=n*R*T./(v-n*b)-n^2*a./v.^2;
dvdp=diff(v)./diff(p);
beta_T=-1./v(2:end).*dvdp;% v begins from the 2nd term as vector dvdp is one term shorter v
disp('   v       p      Beta_T')
fprintf('%7.4f%7.4f %7.4f\n',[v(2:end);p(2:end);beta_T])
plot(p(2:end),beta_T)            % p begins from the 2nd term as vector beta_T is one term shorter p
grid on
xlabel('Pressure, atm')
ylabel('Coefficient \beta_T,atm^{-1}')
title('Isothermal compressibility coefficient')
```

The running command, outputted values of the isothermal compressibility coefficient and generated $\beta_T(p)$ plot are:

```
>> ApExample7(3.5924,0.04271,2,293.15)
   v       p      Beta_T
 9.0000  5.2196  0.1112
13.0000  3.6404  0.1948
```

```
17.0000  2.7947  0.2782
21.0000  2.2678  0.3615
25.0000  1.9081  0.4448
29.0000  1.6468  0.5280
33.0000  1.4485  0.6112
37.0000  1.2928  0.6944
```

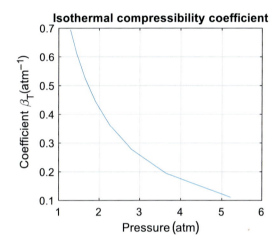

4.6 Questions and exercises for self-testing

1. The first line of the user-defined function should contain: (a)—the function definition; (b)—help comments; (c)—definition of global variables; (d)—both the function definition and the help comments. Choose the correct answer.
2. Local variables are: (a) the variables defined as input/output arguments in the function definition line; (b) all variables used within the function file only; (c) the variables appearing in the workspace after the function file has completed its work; (d) the (b) and (c) answers are correct. Choose the correct answer.
3. Choose the input and output arguments of the following function definition command function y=Ch5_ApE5(lam1,lam2,tf). (a) lam1, lam2, tf—input arguments and y—output; (b) lam1, lam2, tf—output, y—input; (c) lam1, lam2, tf—input and Ch5_ApE5 and y—output: (d) Ch5_ApE5, lam1, lam2, tf—input, y—output.
4. The unterp1 function can be used: (a) for interpolation only; (b) for extrapolation only; (c) for integration; (d) both for interpolation and extrapolation. Choose the correct answer.
5. To find x in the equation $ae^{-kx+c} = b$ with the aid of the fzero function, the equation should be represented as: (a) $ae^{-kx+c} - b = 0$; (b) $\ln(b/a) = -kx + c$; (c) $\ln(b/a) = -kx + c$; (d) just as it is written: $ae^{-kx+c} = b$. Choose all the correct answers.

6. The function $f(x)$ is given by data points; which MATLAB® function should be used for its integration—(a) `trapz`, (b) `fzero`, (c) `quad`, (d) `interp1`. Choose all the correct answers.

7. The function that can be used for defining the maximum of a function called as: (a) `diff`; (b) `quad`; (c) `fminbnd`; (d) `trapz`. Choose all the correct answers.

8. For given vectors of the x and y values, the dy/dx derivative can be calculated as: (a) `diff(y./x)`; (b) `diff(y)./diff(x)`; (c) `diff(y,1)./diff(x,1)`; (d) `diff(y)/ diff(x)`. Choose the correct answer.

9. To find minimum for the function $5x^{3.1} = 4$ with the `fminbnd` function, the following command/s should be written: (a) `x= fminbnd('5*x.^3.1-4',3)`; (b) `x= fminbnd(@(x)'5*x.^3.1-4',3)`; (c) `x= fminbnd('5*x.^3.1=4',3)`; (d) `x= fminbnd(@(x)'5*x.^3.1=4',3)`. Choose all the correct answers.

10. To calculate integral the $f(x)$ function given as a table the following command should be used: (a) `quad`; (b) `quadl`; (c) `trapz`; (d) `integral`. Choose all the correct answers.

11. Concentrations of the substances in solutions (mostly in water) are measured in various units, such as ρ in mg/m^3 and M in ppm (parts per million). The following expressions convert ρ to M (based on 1 atm and 25°C):

$$M = \frac{24.45\rho}{M_1}$$

where M_1 is the molecular weight of the substance, g/mol.

Write and save a script file that calculates and displays the concentration in ppm by the requested (with the `input` command) values of ρ and molar mass M_1 in the above units. Calculate the concentration in ppm for $\rho = 1.5 \cdot 10^3$ mg/m^3 and $M_1 = 39.997$ g/mol (sodium hydroxide).

12. The air temperature T_d when water vapor begins to condense out of the air (dew point), at actual relative humidity RH, can be approximately calculated with the following set of equations:

$$e = 6.112 e^{\frac{17.67 T_w}{T_w + 243.5}} - p(T - T_w)0.00066(1 + 0.00115 T_w)$$

$$RH = 100 \frac{e}{6.112 e^{\frac{17.67 T}{T + 243.5}}}$$

$$T_d = \frac{243.5 \ln \frac{e}{6.112}}{17.67 - \ln \frac{e}{6.112}}$$

where T is the actual, dry-bulb, air temperature in °C; T_w is the wet-bulb temperature (thermometer covered by water-socked cloth) in °C; p is barometric pressure in mbar; RH is relative humidity in %.

Write and save a script file that calculates and displays T_d and RH values for the requested (with the input command) T and p. Take, for example, the $T=25$ and $T_w=19°C$ at $p=985.1$ mbar; use the fprintf command that displays T and T_d with 1 decimal digit, and RH as integer.

13. Dissociation of a substance is described by the equation $m=m_0 e^{-kt}$ where m_0 and m are the substance amounts at the initial and final time respectively; k is dissociation constant. Write and save the user-defined function (with name Problem_4_13) that calculates the dissociation rate $r=-dm/dt$ at $t=0, 0.1, \ldots, 3$ min, $k=1.8$ min^{-1}, and $m_0=5$ mg, and plots the curves $m(t)$ and $r(t)$ in the same graph. Make m_0, k, t_{start}, t_{end}, and t_{step} (where t_{start}—initial, t_{end}—final, and t_{step}—step values of t) as the input and r as the output parameters of the function.

14. Resistivity ρ, Ω m, of a metallic conductor as function of temperature T, K, is presented in the table.

T (K)	300	400	500	600	700	800	900	1000
ρ (mΩ mm)	0.017230	0.02400	0.03088	0.03790	0.04512	0.05260	0.06039	0.06859

Write and save the user-defined function with name Problem_4_14 that can interpolate and extrapolate the data. Make the temperature/s for inter/extrapolation (named Ti), and calculated resistivity (named roi) as the input and output of the function parameters respectively. Use the linear method of the interp1 function. Calculate resistivity at Ti=293 and 550 K. Generate the tabular (marked as circles) and interpolated (marked as stars) points in the same plot.

15. The Lennard-Jones-type intermolecular pair potential in the dimensionless (reduced) form is:

$$U = 4\left[\left(\frac{1}{r^*}\right)^{2n} - \left(\frac{1}{r^*}\right)^n\right]$$

where r^* and U is dimensionless intermolecular distance and potential energy respectively; n is the integer number.

Write and save the user-defined function that search the minimal distance r^*_{min} for $n=2$ and 6. Give the name Problem_4_15 for the function, and assign vector n as the input and vector r^*_{min} as the output parameters respectively; take the r^*-range from 1 to 5. Generate $U(r^*)$ curve for each n on the same plot, and mark the points of the minimum.

16. The virial equation of state for fluids contains series of the virial coefficients from which the second virial coefficient can be calculated by the equation:

$$B^* = 3\int_0^\infty \left(e^{-\frac{4\left[\left(\frac{1}{r^*}\right)^{2n} - \left(\frac{1}{r^*}\right)^n\right]}{T^*}} - 1\right) r^{*2} dr^*$$

where B^* is the normalized by the hard spheres volume the dimensionless second virial coefficient for Lennard-Jones-type potential function (see preceding problem); T^* and r^*—dimensionless temperature and intermolecular distance respectively, n—integer.

Write and save the user-defined function with name `Problem_4_16`, and with vector T^* and number n as input and B^*-value as output function parameters. The function should calculate the B^* for $T^* = 0.5, 1, 5, 10, 15$, and 20 with $n = 6$. Use the `quadgk` command for calculating the semiinfinite integral of B^*.

17. The compressibility factor Z via the second virial coefficient has the approximate form:

$$Z = 1 + B^* p$$

here p is the dimensionless pressure, and B^* is the dimensionless second virial coefficient of the Lennard-Jones potential.

Write and save the user-defined function with name `Problem_4_17`. The function should have the input parameters only, i.e., vector T^* and integer number n (in accordance to the $2n$-n Lennard-Jones potential—see preceding problem), and generates plot with the obtained compressibility factor as function of the pressures $p = 0, 0.01, \ldots, 0.07$ at three isotherms $T^* = 0.5, 1$, and 2. The values of B^* should be received using the `Problem_4_16` function (see problem 16) with the three-isotherm T^* values of and parameter $n = 6$ as the input of this function.

18. The Boyle temperature, arise in material thermodynamics, determines as the temperature for which the second virial coefficient $B^*(T^*)$ becomes zero. Write and save the user-defined function with name `Problem_4_18`. The function should have the dimensionless temperature T^* and number n of the LJ-type potential as the input parameters and the dimensionless Boyle temperature `T_Boyle` as the output parameter. The second virial coefficient should be obtained by the call to the `Problem_4_16` with the inputting temperatures $T^* = 1, 2, \ldots, 20$ and parameter $n = 6$; calculate by the defined second virial coefficients, the Boyle temperature using the `interp1` function when the interpolated B^* value equals to zero. Generate the plot that shows the solid line of the second virial coefficient as a function of temperature and the Boyle point marked with circle. Add the formatting commands to the plot.

19. One of the equations that relates temperature T (K), pressure P (atm) and volume V (L/mol) of the substance is the Redlich-Kwong equation:

$$P = \frac{RT}{V-b} - \frac{a}{V(V-b)\sqrt{T}}$$

where R is the gas constant equals to 0.08206 atm/(mol K), a and b are constants that can be calculated by the expressions:

$$a = \frac{0.42748 R^2 T_c^{\frac{5}{2}}}{P_c} \quad \text{and} \quad b = \frac{0.08664 RT_c}{P_c}$$

where T_c and P_c are temperature and pressure at the critical point of a substance.

Write the script program with the `Problem_4_18` name. The program should calculate the volumes V at temperatures $T=273.15$, 293.15, 313.15, and 333.15 K, $P=2$ atm, $T_c=126.2$ K, and $P_c=33.5$ atm (Nitrogen). Use for defining V the `fzero` function with the initial volume $V_0=22$ L/mol. Display results with the `fprintf` function as a two-column table, the first column T and the second V with two decimal digits for each value.

20. To estimate the fugacity coefficient of a gaseous or fluidic substance through the Redlich-Kwong compressibility factor Z, the following integral should be calculated:

$$I = \int_{0.0001}^{P} \frac{Z-1}{P} dP$$

where P is the pressure in atm, $Z = \frac{1}{1-h} - \frac{A^2}{B} \frac{h}{1+h}$, $h = \frac{0.08664 RT_c}{P_c V}$, $A^2 = \frac{0.42748 T_c^{\frac{5}{2}}}{P_c T^{\frac{5}{2}}}$, $B = \frac{0.08664 T_c}{P_c T}$,

$R = 0.08206$ atm/(mol K), V—volume in L/mol, T—temperature in K, T_c and P_c are respectively the temperature and pressure at the critical point of a substance. The volume V in the h-expression is defined from the equation $P = \frac{RT}{V-b} - \frac{a}{V(V-b)\sqrt{T}}$ with $a = \frac{0.42748 R^2 T_c^{\frac{5}{2}}}{P_c}$ and $b = \frac{0.08664 RT_c}{P_c}$.

Write and save the user-defined function with name Problem_4_20 that has the T vector and single numbers P, P_c, and T_c as the input parameters, and I as output parameter. The volume V should be defined by the `fzero` function with $V_0 = 22$ L/mol and the integral I with the `integral` function. The input parameters are $T=293.15$, 213.15, 333.15, $P=2$; $T_c = 154.6$, and $P_c = 49.8$ (Oxygen).

21. Radioactive substances are decomposed with time; the process can be described by the decay equation:

$$N = N_0 \left(\frac{1}{2}\right)^t$$

where t is the number of days of half-life periods elapsed (the half-life being the time it takes for a decaying substance to disintegrate by half); N_0 is the initial amount of the substance, and N is the residual amount both in μCi (microCuries).

Write the script program with the `Problem_4_21` name. The program should calculate the time t when the radioactivity N is equal to $N_0/4$, $N_0/10$, $N_0/100$, and $N_0=500$ μCi. Enter the N_0 and N values with the two `input` commands, the first—enters the single number 500 and the second—enters the following vector [500/4 500/8 500/16]. Use the `fzero` function with the initial t equal to 5. Display results with the `fprintf` function that outputs two-column matrix with N as the first column and t the second.

4.7 Answers to selected questions and exercises

2. (b) all variables used within the function file only.
4. (d) both for interpolation and extrapolation.
7. (c) fminbnd.
9. (a) x=fminbnd('5*x.^3.1-4',3) and (b) x=fminbnd(@(x)'5*x.^3.1-4',3).
10. (c) trapz.
12.

```
>>Problem_4_12
Enter actual air temperature in C,T= 25
Enter wet-bulb air temperature in C,T= 19
Enter barometric pressure in mbar,p= 985.1
Temperature, C 25.0 Humidity, 57 Dew point, 15.8
```

14.

```
>> roi=Problem_4_14([293 550])
roi =
    0.0168  0.0344
```

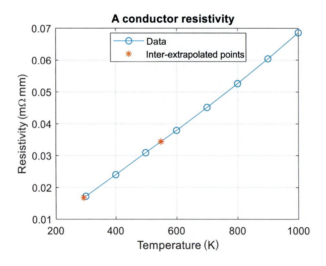

16.

```
>>B=Problem_4_16(6,[0.5 1 5:20])
B =
  Columns 1 through 6
   -8.7202 -2.5381  0.2433  0.3229  0.3761  0.4134
```

```
Columns 7 through 12
 0.4406 0.4609 0.4763 0.4882 0.4975 0.5048
Columns 13 through 18
 0.5106 0.5151 0.5187 0.5215 0.5237 0.5254
```

18.
```
>> T_boyle=Problem_4_18(6,1:20)
T_boyle =
   3.4170
```

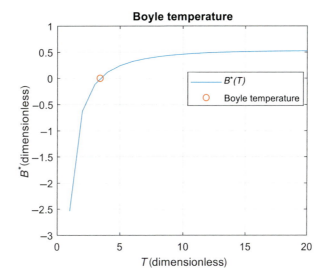

20.
```
>> I=Problem_4_20([293.15:20:333.15],2,154.6,49.8)
I =
  -0.0161 -0.0120 -0.00
```

5

Curve fitting commands and the Basic Fitting tool

In order to analyze data, a materials specialist must frequently match them with some empirical or theoretical expression. If some mathematical relation describes data accurately that allows the prediction of, for example, a substance behavior and select a material with the better properties, or to evaluate the discrepancies between different series of measurement values and define error of the most accurate data, etc.

In this chapter, we show the governor commands and specialized tool which materials scientist/engineer can use to describe the data with a mathematical expression. The use of polynomial and other relationships binding two variables together is described. In addition, how to use an optimization command to fit the relation between more than two variables is explained.

5.1 Fitting with polynomials and some other functions

Finding the best match for some expression or curve to the data points is called curve fitting or regression analysis. The suitable expression may be selected empirically or based on a theory. The curve fitting involving polynomials, and some nonpolynomial functions are described below.

5.1.1 Fitting by polynomials

The process concludes in finding the values of the coefficients of a mathematical expression, which should fit the experimental data. The polynomial expression for the fit is:

$$y = a_{n-1}x^{n-1} + a_{n-2}x^{n-2} + \ldots + a_1 x + a_0$$

where x is the variable that should be connected with another variable $-y$; n, and a_i, are the number of items and coefficients of polynomial respectively; i is the serial number of the a.

The highest value of the exponent, $n-1$, is called the degree of the polynomial, e.g., the straight line $y = a_1 x + a_0$ is first-degree polynomial that has two coefficients, a_1 and a_0, to be determined by the given pairs of the x and y values denote the point for fit.

The n coefficients of a_i can be determined by the n of (x,y) points solving the set of $y(x)$ equations written for each of the points. These equations are linear in respect to the coefficients a_i. The fitting when the number of coefficients is the same as that of points is not always efficacious. Although such a fit gives a complete match with the data points, however, it often leads to significant discrepancies in spaces between the data points. The linear $y(x)$ dependence is another case of possible inefficiency of such procedure since two points suffice to determine the a-coefficients, but experimental data contain a larger number of points, which means more equations than unknowns.

The most commonly used method for better fit is the least squares method, in which the coefficients are determined by minimizing the sum of squares of the differences between the fitting equation and data values; these differences are called residuals and denoted R. The minimization is performed by taking the partial derivative of the squared R with respect to each coefficient and equating the derivatives to zero, e.g., for the straight line $\frac{\partial \sum R_i^2}{\partial a_1} = 0$ and $\frac{\partial \sum R_i^2}{\partial a_0} = 0$, and the set of these two equations should be solved for obtaining the a_1 and a_2 values (see example in Section 2.3.4.1).

For fitting by a polynomial, MATLAB® has the `polyfit` function, whose simplest form is:

```
a=polyfit(x,y,m)
```

where the input arguments x and y are the vectors of the (x,y)-point values, and m is the degree of the polynomial; the output argument a is the $(m+1)$-element vector of the fitting coefficients. The first element a(1) is the a_{n-1} coefficient, the second a(2) is a_{n-2}, ..., and the final a(m+1) is a_0.

For example, the 1st degree polynomial with the defined a coefficients looks as y=a(1)*x+a(2), the second degree polynomial as y=a(1)*x²+a(2)*x+a(3), and so on.

The y-values can be calculated by the fitting coefficients at any x of the fitted interval. In MATLAB®, it can be done with the `polyval` command; which has the following simplest form:

```
y_poly=polyval(a,x_poly)
```

where a is the vector of polynomial coefficients as per the `polyfit`, x_poly is that of the coordinates at which the y_poly values are calculated.

For example, the relative dielectric constant ε of an organic compound is 1, 1.045, 1.074, 1.129, and 1.163 (dimensionless) at pressures $P=0.1$, 50, 100, 250, and 400 (MPa) respectively.

Problem: Write script file with name Ch5_PoyfitEx that fits the $\varepsilon(P)$ data by the first-degree ($m=1$) and second degree ($m=2$) polynomials, and generate two fit curves together with the data in the same plot.

The following script file, named Ch5_PoyfitEx, solves the problem.

```
% Curve fitting by polynomial
% x – pressure, MPa; y – dielectric constant, dimensionless
x=[0.1, 50, 100, 250, 400];                          % vector of x, pressures
y=[1, 1.045, 1.074, 1.129, 1.163];                   %vector of y, dielectric constant
format shortG                                         % best format
a_linear=polyfit(x,y,1)                              % coefficients for 1st-degree polynomial
a_quadratic=polyfit(x,y,2)                           % coefficients for 2nd-degree polynomial
x_pol=linspace(x(1),x(end));                         % vector of 100 x values for graph
y_linear =polyval(a_linear,x_pol);                   % y values by 1st-degree polynomial
y_quadratic =polyval(a_quadratic,x_pol);             % y values by 2nd-degree polynomial
plot(x,y,'o',x_pol,y_linear,'-',x_pol,y_quadratic,'--')          %data and polynomials
xlabel('Pressures, MPa'), ylabel('Dielectric constant, nondim'), grid on
legend('Original data',' 1st-degree polynomial fit',' 2nd-degree polynomial
fit','location','best')
format                                               %returns to the default format
```

The `a_linear` and `a_quadratic` variables are the vectors containing the coefficients defined by the `polyfit` commands for the 1st- and 2nd-degree polynomials respectively; defined coefficients are displayed with the `format shortG` command (that is the best for displaying the values of very different orders). The `y_linear` and `y_quadratic` variables are the vectors with dielectric constants obtained by the defined coefficients of the above-named polynomials. The data (circled points) and two fitting curves (solid and dashed lines) generated with the `plot` command; the `legend` command explains the plotted data and series. The final format command sets the default display format.

After typing and entering the file name in the Command Window, the following coefficients are displayed, and the graph is generated:

```
>> Ch5_PoyfitEx
a_linear =
 0.00038639  1.0204
a_square =
-8.5247e-07 0.00073024   1.0054
```

In accordance to the coefficients `a_linear` and `a_quadratic`, the fit polynomials for the dielectric constant are:

$$\epsilon_1 = 3.8639 \cdot 10^{-4} P + 1.0204$$

$$\epsilon_2 = -8.5247 \cdot 10^{-7} P^2 + 7.0324 \cdot 10^{-4} P + 1.0054$$

where ϵ_1 and ϵ_2 are relative dielectric constants fitted with the 1st- and 2nd-degree polynomials, and P is pressure, in MPa.

5.1.2 Fitting with nonpolynomial functions

Many materials science applications require an exponential, power, or a logarithmic fitting function:

$$y = a_0 e^{a_1 x}$$

$$y = a_0 x^{a_1}$$

$$y = a_1 \ln x + a_0 \quad \text{or} \quad y = a_1 \log x + a_0.$$

Each of these functions can be fitted by polynomial with the `polyfit` command. For this, they should be transformed to the linear form $y = a_1 x + a_0$. Two functions with logarithms are already in this form, as $\ln x$ or $\log x$ are apparently the modified x. The first two functions can be performed in this form by taking the logarithm of both parts of the equations, that gives:

$$\ln y = a_1 x + \ln a_0$$

$$\ln y = a_1 \ln x + \ln a_0$$

Representation of these equations with the `polyfit` functions, and relations between coefficients a_1, a_2 of the original equations and coefficients a defined by the `polyfit` are as follows:

Equation	The `polyfit` function	The a_0 and a_1 via vector a
$y = a_0 e^{a_1 x}$	`a=polyfit(x,log(y),1)`	$a_1 = a(1)$, $a_0 = \exp(a(2))$
$y = a_0 x^{a_1}$	`a=polyfit(log(x),log(y),1)`	$a_1 = a(1)$, $a_0 = \exp(a(2))$
$y = a_1 \ln x + a_0$	`a=polyfit(log(x),y,1)`	$a_1 = a(1)$, $a_0 = a(2)$
$y = a_1 \log x + a_0$	`a=polyfit(log10(x),y,1)`	$a_1 = a(1)$, $a_0 = a(2)$

To select the function for fitting, it is useful to estimate its suitability to fit the data. For this, the data should be built in the certain axes—linear, logarithmic, semilogarithmic. The first two parameters in the `polyfit` function represent these axes scaling, e.g., for the $y = a_0 x^{a_1}$ function (represented by the `polyfit(x,log(y),1)` command) the x and y axes should be logarithmic, and for the $y = a_0 e^{a_1 x}$ function (the `polyfit(x,log(y),1)` command) the x axis is linear and y-logarithmic.

Consider, for example, the fit with the power $y = a_0 x^{a_1}$ function the speed of sound—temperature data. The c-values for a fluid equal to 1447, 1481, 1526, 1552, and 1556 m/s were measured at temperatures $T = 10, 20, 40, 60,$ and $70°C$ respectively.

Problem: Fit these data by the `polyfit` commands, and generate two plots at the same Figure window: the log*c*-log*T* curve (solid line with the circled points) and the *c*(*T*) curve (solid line) together with data (circles) in the same plot.

The following commands, saved as the `Ch5_PowerfitEx` script file, solve the problem.

```
% Curve fitting for the power function
% T – temperature, C
% c – speed of sound, m/sec
T=[10 20:20:60 70];                        % vector of temperatures
c=[1447 1481 1526 1552 1556];              % vector of speed of sound
a=polyfit(log(T),log(c),1);% coefficients for T=ao*T^(a1)by 1st degree polynomial
Tg=10:5:70;                                % vector of T values for graph - Tg
a0=exp(a(2))                               % a0 via defined vector of coefficients
a1=a(1)                                    % a1 via defined vector of coefficients
cg=a0*Tg.^a1;                              % calculates c by the power function
subplot(1,2,1)
loglog(T,c,'-c'), grid on                  % c(T) data in logarithmic coordinates
xlabel('Temperature, C'),ylabel('Speed of sound, m/sec')
title('Power function in logarithmical coordinates')
subplot(1,2,2)
plot(T,c,'o',Tg,cg),grid on                % c(T) and original data
xlabel('Temperature, C'),ylabel('Speed of sound, m/sec')
title('Fit with the power function')
legend('Original data','fitting with power function')
```

In these commands, parameter `a` is the vector containing the coefficients defined by the `polyfit` commands. The variables `a0` and `a1` are the coefficients of the power equation calculated with the `a`-vector; obtained values are displayed. The `Tg` and `cg` variables are the vectors with temperatures and sound speeds obtained by the power function with the coefficients `a0` and `a1`; these vectors are used in the `plot` command for generating the fitting curve in the graph. The `loglog` command plots data in logarithmic coordinates to check whether log*T*-log*c* curve is close to a straight line. The `plot` command generates data points together with the fitting curve to demonstrate the results.

After typing and entering the file name in the Command Window, the following coefficients are displayed, and the graph is generated:

```
>> Ch5_PowerfitEx
a0 =
  1.3222e+03
a1 =
  0.0387
```

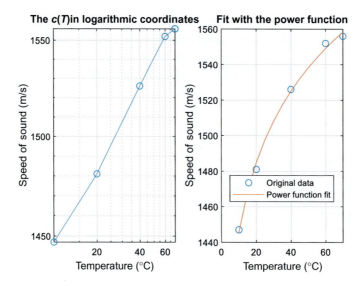

5.1.3 About the goodness of fit

The discrepancy between the data and an equation that fits the data characterizes the goodness of fit. There is not one criterion for this purpose. One of the criteria that is widely used is the so-called coefficient of determination or R-squared:

$$R^2 = 1 - \frac{\sum_{i=1}^{n}(y_i - y_{\text{fit}\,i})^2}{\sum_{i=1}^{n}(y_i - \bar{y})^2}$$

where y and y_{fit} are respectively the values that were observed and predicted with the modeled equation; $\bar{y}$ is the average of the observed values; n—amount of the i-points to be fitted.

This criterion varies from zero to one. A fit is considered best if R-squared is closest to 1.

Calculate, for instance, the R-squared for the 1st and 2nd degrees polynomials defined by the Ch5_PolyfitEx program (example of the Section 5.1.1). It can be done by the addition to this program of the following commands:

```
% R squared for the 1st- and 2nd-degree polynomials

y_lin=polyval(a_linear,x);      % predicted y-values for the 1st-degree polynomial
y_qua=polyval(a_quadratic,x);   % predicted y-values for the 2nd-degree polynomial
s1_lin=sum((y-y_lin).^2);       % squared residuals for the 1st-degree polynomial
s1_qua=sum((y-y_qua).^2);       % squared residuals for the 2nd-degree polynomial
s2=sum((y-mean(y)).^2);
R2_linear=1-s1_lin/s2           % R-squared for the 1st-degree polynomial
R2_quadratic=1-s1_qua/s2        % R-squared for the 2nd-degree polynomial
```

After running the Ch5_PolyfitEx program modified with this commands, the *R*-squared values for the 1st- and 2nd-degree polynomials are displayed in addition to the defined coefficients of the polynomials:

```
>> Ch5_PoyfitEx
a_linear =
  0.00038639  1.0204
a_quadratic =
  -8.5247e-07 0.00073024  1.0054
R2_linear =
  0.9435
R2_quadratic =
  0.9935
```

As can be seen, the *R*-squared for 2nd-degree polynomial is closest to 1 then for the 1st-degree, thus the quadratic polynomial better depicts the original data on the dielectric constant. The same conclusion can be made from the graphical representation of the original points and fitting curves for both polynomials to be fitted (Fig. 5.1).

Another frequently used criterion of the goodness of fit is the norm of residuals:

$$R_{\text{norm}} = \sqrt{\sum_{i=1}^{n}(y_i - y_{\text{fit}_i})^2}$$

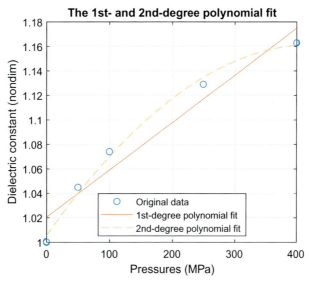

FIG. 5.1 Fitting with the first- and second-degree polynomials.

where all notifications have the same sense as for the *R*-squared above; nevertheless, contrary to the *R*-squared, the lower value of this criterion reflects a better fit.

Continuing the example above, we can calculate the norm of residuals for the 1st and 2nd degree polynomials by the addition the following commands:

```
Rnorm_linear=sqrt(s1_lin)        % R_norm for the 1st-degree polynomial
Rnorm_quadratic=sqrt(s1_qua)     % R_norm for the 2nd-degree polynomial
```

After running the Ch5_PolyfitEx program modified now with these commands, the following values are displayed:

```
>> Ch5_PoyfitEx
a_linear =
  0.00038639   1.0204
a_quadratic =
  -8.5247e-07 0.00073024   1.0054
R2_linear =
  0.9435
R2_quadratic =
  0.9935
Rnorm_linear =
  0.0309
Rnorm_quadratic =
  0.0105
```

The `Rnorm_quadratic` value is significantly smaller then `Rnorm_linear`; therefore, 2nd order fit is better than linear. It is obvious that the R_{norm} criterion brings the same conclusion as the *R*-squared.

Note: These, like other available criterions, are imperfect since they can show the best fit for polynomials that pass exactly through the original points but oscillate between them. Thus, it is recommended to plot the original points along with the fitting curve to ensure the goodness of fit.

5.2 Interactive fitting with the Basic Fitting interface

In addition to the polynomial fitting commands described above, MATLAB® has a specific tool for interactive fit called Basic Fitting; the data in this tool can also be interpolated. The tool permits:

- Polynomial fit the data points with the linear, quadratic, cubic and up to 10th degree polynomials;
- Showing the defined polynomial equations and plotting one or more fit curves for better comparison;

Chapter 5 • Curve fitting commands and the Basic Fitting tool

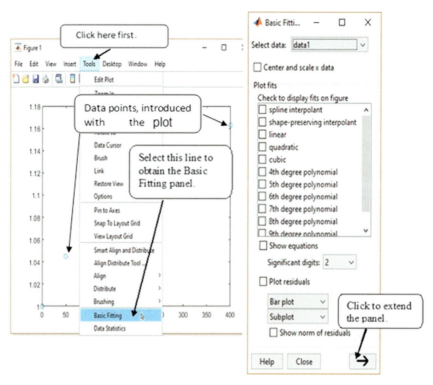

FIG. 5.2 Activating the Basic Fitting interface; Figure window with the popup menu having the Basic Fitting line to start (to the left) and first the "Plot fits" panel of the Basic Fitting window.

- Plotting the residuals of the each polynomial fit together with the norm of residual values;
- Predicting the values by the fit polynomials at various points input by the user.

To open the tool, it is first necessary to plot the data points to be fitted. After the plot appears, select the Basic Fitting line from the Tools option of the Figure window menu (see the left figure in Fig. 5.2). Immediately after this action, the "Plot fits" panel of the Basic Fitting window appears (see the right figure below).

To extend the "Plot fits" panel, click the "Show next panel" button → . The "Numerical results" panel appears. Next, click on this button opens the next panel and finally the entire Basic Fitting window appears (Fig. 5.3). To return back to the previous panel, the "Hide last button back" button ← should be clicked.

The following explains the items located in each panel of the Basic Fitting Window.

5.2.1 The "Plot fits" panel items

5.2.1.1 "Select data" field
Only one data set presented in the Figure window should be chosen for the fit by this item.

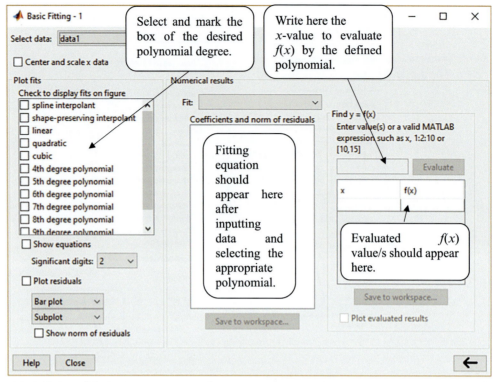

FIG. 5.3 Final view of the Basic Fitting window after two repeated extensions.

5.2.1.2 "Center and scale x data" box

For a better fitting, the x-values of the data points can be centered and scaled by marking this box. Each x value, i, recalculates with the equation $z_i = (x_i - \mu)/\sigma$, in which μ is the x values average, σ is the standard deviation of x-values and z is the x value after centering and scaling. Frequently, when this option is not marked but the high-degree polynomial is selected to fit or/and x-values are very small or very large, the special nameplate appears with a message about poor conditionality of the polynomial with advice to center and scale the data.

5.2.1.3 "Check to display fits on figure" area

This item allows the user to select fit by polynomials from the 1st (linear) degree to the 10th degree and interpolation by the spline or Hermite method. More than one fit box can be marked in this item. When the polynomial degree was selected, the fit line appears immediately in the plot and the polynomial equation with coefficients and norm of residuals appear in the "Coefficients and norm of residuals" item of the "Numerical results" panel (see below).

5.2.1.4 "Show equation" box and "Significant digits" field
The first of these items to be marked displays the fit equation on the plot. The second item allows the user to choose the number of significant digits displayed within the equation in the plot.

5.2.1.5 "Plot residuals" box with two adjacent fields
To plot residuals, mark this item. Residuals R_i calculate with the equation $y_i - y_{\text{fit}_i}$, where the original y_i and the calculated y_{fit_i} are values at each i-point. The first of the adjacent fields enabled to select the resulting plot type: bar, scatter, or line. The second permits to select the way to plot the residuals—in separate Figure window or in the same window where the dataset and fitted curve were plotted.

5.2.1.6 "Show norm of residuals" box
This item after marking displays a good fit by the norm of residuals. This norm is calculated as $\sqrt{\sum_{i=1}^{n} R_i^2}$ correspondingly to the Section 5.1.3.

5.2.2 The "Numerical results" panel items

5.2.2.1 "Fit" dropdown list
This item is the dropdown menu that helps to explore, independently from the previous choices, a number of fits and to see the fit equation, coefficients, and residual norm for one of the selected fits. Requested data are displayed in the "Coefficients and norm of residuals" box.

5.2.2.2 "Coefficients and norm of residuals" box
This item displays the selected polynomial, defined coefficient values, and norm of residuals.

5.2.2.3 "Save to workspace" button
After clicking this button, the window appears with boxes for marking to save the fit, residuals, and norm of residuals in the MATLAB® workspace. The default names of these parameters can be changed by the user.

5.2.3 The "Find y=f(x)" panel items

5.2.3.1 "Enter value(s) or a valid MATLAB® expression…"[a] area
The any values of x can be typed in the field under this heading line: to calculate corresponding y values by the defined fit expression. The calculated y-values will display in the box below after clicking the "Evaluate" button located to the right of this item field.

---

[a] The caption of the item is shortened.

5.2.3.2 "Save to workspace" button
The small windows appear after clicking this button that provides the choice of the variables and its names to be saved in the MATLAB® workspace.

5.2.3.3 "Plot evaluated results" box
When this item is marked, the evaluated y-values are displayed in the plot.

5.2.4 An example of using the "Basic Fitting" tool

Let us take, for example, the data on the dielectric constant ε (dimensionless) as a function of pressure P (in MPa), which were fitted in Section 5.1.3 using the `polyfit` command. Here this data is used for the same purpose with the Basic Fitting interface.

Problem: Define linear and square polynomial equations that describe the $\varepsilon(P)$ dataset; show fitted coefficients and norm of residuals in the Fit panel and on the plot, evaluate dielectric constant at pressure $P = 200$ MPa by the fit equation with the lesser norm of residuals; save defined equations, norm of residuals, and evaluated dose value in the workspace.

To solve the problem with the Basic Fitting interface, first select the dataset in Command Window and plot the data. The following commands realize this:

```
>> x=[0.1, 50, 100, 250, 400];          % vector of x, pressures
>> y=[1, 1.045, 1.074, 1.129, 1.163];   %vector of y, dielectric constant
>> plot(x, y, 'o')                       % plots data points as circles
```

After the Figure window appears with the plot, select the Basic Fitting line from the Tools menu in this window. This activates the Basic Fitting window that should be extended in full with the "Show next panel" button as described above. Mark the `linear` box in the "Plot fits panel," the defined linear equation `y=p1*x+p2`, coefficients `p1=0.00038639` and `p2=1.0204`, and norm of residuals `0.030927` appear into the "Coefficients and norm of residuals" box of the "Numerical results" panel appears.

Check then the following boxes: "Show equations," "Plot Residuals," and "Show norm of residuals." Two subplots appear in the Figure Window—the fit curve with the typed linear equation and original data, and the residual bar graph with residual norm value. Now select the "quadratic" box in the "Plot fits" panel. The defined quadratic equation `y=p1*x^2+p2*x+p3`, coefficients `p1=-8.5247·10^-7`, `p2=-0.00073024`, `p3=1.0054` and norm of residuals `0.010483` appear in the "Coefficients and norm of residuals" box of the "Numerical results" panel. These results are the same as the result obtained in Sections 5.1.1 and 5.1.3.

Both the linear and cubic polynomial curves together with the fit equations (default is coefficients with two significant digits) appear on the upper subplot of the Figure window. Simultaneously, the residual bars with their norms appear on the lower subplot. The norm of residuals of the quadratic polynomial is almost three times lower than the norm of residuals for the linear polynomial. Thus, this polynomial better describes the data and is therefore used for evaluation of the expected dielectric constant at the 200 MPa pressure. Enter the number 200 in the "Enter value(s) …" field and press the "Evaluate" button.

To mark the evaluated point in the upper plot of the Figure window, check the "Plot evaluate results" box in the bottom of the "Find …" panel.

The next two figures show the Basic Fitting tool window for quadratic fit and evaluation and the Figure window with both linear and quadratic fits.

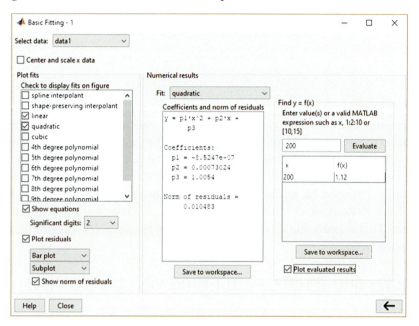

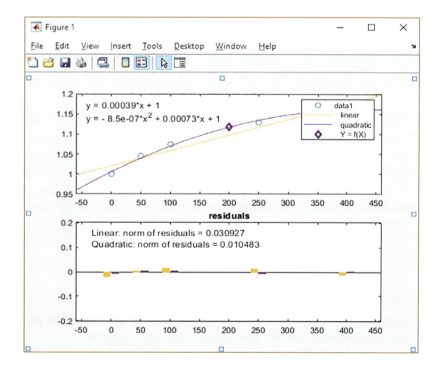

To export fits into the Workspace press now the "Save to Workspace" button of the "Numerical results" panel. The dialog box appears.

After clicking "OK," the `fit`, `normresid`, and `resids` variables are exported and appear in the Workspace window. The `fit` variable has a data type called structure; this concept lies out of the scope of this book. To receive the fit coefficients from the `fit` structure, the `fit.coeff` variable name should be entered in the Command Window. The same operations should be performed to save the evaluated dielectric constant value: press the "Save to Workspace" button in the "Find …" panel; click OK in the dialog box that appears. This exports the x (the pressure) and $f(x)$ (dielectric constant) variables to the Workspace window.

5.3 Single- and multivariate fitting via optimization

Hitherto, the fitting with the polynomials or with resulting to polynomial single-variable functions was discussed. In this section, the nonlinear multivariate fit (also suitable for single) is described; this fit includes also the single-variate fit. Such fit may be done with the optimization command that searches minimum of multivariate function by minimizing the sum of the square errors:

$$\underset{a}{\text{minimize}} \sum_{i=1}^{n} \left[ydata_i - y_{\text{fit}\,i}(a, x1data_i, x2data_i, \ldots) \right]^2$$

where y_i and $y_{\text{fit}\,i}$ are respectively the given (observed in the experiment or taken from the table) and fitting functions; of the $x1data_i$, $x2data_i$, … parameter values at which the $ydata$ were observed; n—number of points, i—current point, a is the vector of fitting coefficients that should be defined.

The many optimization functions belongs to the Optimization toolbox, but in this primer, we represent the basic MATLAB® commands only. Such an optimization command that can be used for fitting is the `fminsearch` command, and one of its forms are:

```
a=fminsearch(@fun(a)(a,x1data,x2data,...,ydata),a0)
```

where

    `ydata` is the vector of the observed values;
    `x1data, x2data, …` are the one, two or more vectors of the observing parameters; these vectors should have the same length as the `ydata` vector;

`@fun(a)(a,x1data,x2data,...,ydata)` is the call to a `fun` function that includes the equation to fit; `fun` is the name, e.g., myfun, myfit, or any other.

a is the vector of coefficients presented in the fitting equation which are searched with the optimization procedure by the derivative-free method[b]; defined a-values are transmitted to the output parameter a;

a0 is vector of the initial values of the a coefficients.

As an example, we fit a gaseous substance P,v,T—data with this command and define in such way the coefficients a and b in the van der Waals equation of state:

$$P = \frac{RT}{v-b} - \frac{a}{v(v+b)T}$$

where $R = 82.06$ atm cm^3/(gmol K) is the gas constant, T is the temperature in K, v is the volume in cm^3/(gmol), P is pressure in atm.

Problem: Write the user-defined function with name `Ch5_MultifitEx` with the a0-vector as input and a and b coefficients as output. The function should determine values of the a and b coefficients by the `fminsearch` command and the following data: $T = 283, 313, 375, 283, 313, 375, 283$, and 375; $v = 480, 480, 576, 672, 576, 672, 384$, and 384; $P = 32.7, 42.6, 44.5, 25.7, 36.6, 38.6, 37.6$, and 63.0. The initial values of a and b coefficients are $1 \cdot 10^{-7}$ and 1 respectively. In addition, the function displays the five-column table with the v, T, P—data, calculated P, and absolute value of the discrepancies between the specified and obtained pressure values (termed residuals).

The following commands, which should be written in the Editor window and saved, solve this problem.

```
function [a,b]=Ch5_MultifitEx(a0)
% multivariate fit by the van der Waals equation
% to run: >> [a,b]=Ch5_MultifitEx([1e7,1])
T = [283, 313, 375, 283, 313, 375, 283, 375];
v = [480, 480, 576, 672, 576, 672, 384, 384];
P = [32.7, 42.6, 44.5, 25.7, 36.6, 38.6, 37.6, 63.0];
a_b=fminsearch(@(a)myfun(a,T,v,P),a0);% calls to myfun and obtaining the a and b
a=a_b(1);b=a_b(2);    % assigns defined coefficient to the van der Waals parameters
Pfit=82.06*T./(v-b)-a./(sqrt(T).*v.*(v+b));    % calculates P by the defined a and b
Res=abs(P-Pfit);
fprintf('   T     v     P    Pfit    Residual\n')
fprintf('%7.2f %7.2f %6.2f %6.2f %5.4f\n',[T;v;P;Pfit;Res])
function sse=myfun(a,T,v,P)  % function containing the sum of the squared residuals
sse = sum((P-82.06*T./(v-a(2))+a(1)./(sqrt(T).*v.*(v+a(2)))).^2);
```

The commands of this user-defined function perform the following actions:

- Defines the function name `Ch5_MultifitEx`, the one input vector a0—initial values of the van der Waals coefficients, and two output single variable a and b—the resulted coefficients of fit

---

[b] The method that does not use derivatives or finite differences; studied in the numerical analysis courses.

- inputs the vectors of the temperature, volume, and pressure values, where the first two represents the x1data and x2data arguments and the third—the ydata argument of the fminsearch command as per above;
- realizes the optimization with the fminsearch command for the equation of state written in the myfun subfunction with starting a0 values and parameters a,T, v, and P are transmitted to the myfun, the output a_b vector contents the defined coefficients;
- assigns defined coefficients a_b to the a and b variables as they are the van der Waals coefficients; these values are transmitted to the Ch5_MultifitEx output;
- calculates pressures Pfit by the van der Waals equation of state using the fitted coefficients and given T and v;
- calculates absolute values of the discrepancies (residuals) between the original and computed pressures;
- displays a caption and the five-column table with the two fprintf commands, the last of them formats the T, v, P, and Pfit displaying values as numbers with two decimal digits and the residual—with four decimal digits;
- defines the myfun subfunction that calculates sum of squares error (SSE) by the call from the fminsearch command; the SSE value obtained with the inputted a, T, v, and P values is assigned to the output variable sse.

The running command, displayed table, and fitting coefficients are:

```
>> [a,b]=Ch5_MultifitEx([1e7, 1])
     T        v        P       Pfit    Residual
   283.00  480.00   32.70    33.69     0.9855
   313.00  480.00   42.60    40.05     2.5512
   375.00  576.00   44.50    45.46     0.9619
   283.00  672.00   25.70    26.87     1.1675
   313.00  576.00   36.60    35.09     1.5052
   375.00  672.00   38.60    39.87     1.2661
   283.00  384.00   37.60    38.02     0.4203
   375.00  384.00   63.00    63.03     0.0255

a =
   7.4224e+07
b =
   30.6824
```

This solution is sensitive to the initial point a0; as a whole, the solution is more accurate when the values of the *R*-squared or R_{norm} criterions are better (Section 5.1.3), here the better achieved *a* and *b* numbers are presented. This fit also can be improved with the additional forms of the fminsearch command that consider accuracy, repeatability, amount of iterations, etc. For more detailed study of this command options, type >> doc fminsearch in the Command Window.

5.4 Application examples
5.4.1 Fitting of the compressibility factor with the virial series

The virial-type equation of state has polynomial form to be written for the compressibility factor:

$$Z = A + B\rho + C\rho^2 + D\rho^3 + \ldots$$

where ρ is density of a material, and A, B, C, D, ...—virial coefficients.

Measured at $T = 300$ K, the compressibility factor of a substance as function of density are: $Z = 0.9999$, 0.9950, 0.9917, 0.9901, 0.9903, 1.0326, 1.1089, 1.2073, 1.3163 bar, and $\rho = 0.0401$, 0.8059, 1.6171, 2.4296, 4.0486, 7.7654, 10.8467, 13.2835, 15.2294 mol/L.

Problem: fit the $Z(\rho)$ data with a 3rd degree polynomial and calculates discrepancies (i.e., residuals) between the experimental and calculated by the virial equation values of the compressibility factor. Write the program as user-defined function named ApExample5_1 with the virial coefficients as output parameters and without the input parameters. The program also should generate graph $Z(\rho)$ with the experimental points and fitted curve, in addition the program should display the table with the experimental ρ, Z data, and with calculated Z and the absolute residual values.

The commands solving this problem are:

```
function [A,B,C,D]=ApExample5_1
% fits the compressibility factor with the cubic polynomial
% To run: >>[A,B,C,D]=ApExample5_1
close all
Z=[0.9999 0.9950 0.9917 0.9901 0.9903 1.0326 1.1089 1.2073 1.3163];
ro=[0.0401 0.8059 1.6171 2.4296 4.0486 7.7654 10.8467 13.2835 15.2294];
vir_coeff=polyfit(ro,Z,3);                     % 3rd degree polynomial fit
D=vir_coeff(1);C=vir_coeff(2);B=vir_coeff(3);A=vir_coeff(4);% fitting coefficients
Zfit=polyval(vir_coeff,ro);                    % z by the fitting polynomial
residual=abs(Z-Zfit);                          % residuals: z-z_fit
rog=linspace(ro(1),ro(end));        % density values for the fitting curve plot
Zg=polyval(vir_coeff,rog); % compressibility factor values for the fitting curve plot
plot(ro,Z,'o',rog,zg)                          % resulting graph
grid on
xlabel('Density, mol/L'),ylabel('Compressibility factor')
legend('Data','Fit','location','best')    % legend with its best disposition on graph
axis tight                        % sets axis limits to be equal to the data range
fprintf(' Density  Z    Z_fit Residual\n')    % caption of the resulting table
fprintf('%9.4f %6.4f %6.4f %7.4f\n',[ro;Z;Zfit;residual])   % resulting table
```

The commands of this function act as follows:

- define the `ApExample5_1` function with the output variables A, B, C, and D; this user-defined function written without input arguments;
- explain the purposes and represent the command to run in the help lines;
- assign the compressibility factor and density values to the Z and ro output variables;
- fit the $Z(\rho)$ data by the 3rd degree polynomial with the `polyfit` command;

- calculate Zfit by polynomial with the defined polynomial coefficients using the `polyvalue` command;
- assign vector of densities `rog` for further plotting by the `linspace` command with minimal and maximal densities as the first and the last densities of the `ro` vector; calculate vector of the compressibility factors `zg` by the assigned `rog` with `polyvalue` command;
- plot the $Z(\rho)$ data (circles) and fitting curve (solid line), label the axes and show the legend, that is located in the "better" place;
- display the table caption and four columns with the ρ, Z, Zfit, and Residual values.

The running command, the resulting table, the defining fitting (virial) coefficients, and the resulting plot are:

```
>> [A,B,C,D]=ApExample5_1

   Density      Z        Z_fit    Residual
   0.0401    0.9999    0.9995    0.0004
   0.8059    0.9950    0.9951    0.0001
   1.6171    0.9917    0.9918    0.0001
   2.4296    0.9901    0.9900    0.0001
   4.0486    0.9903    0.9918    0.0015
   7.7654    1.0326    1.0305    0.0021
  10.8467    1.1089    1.1092    0.0003
  13.2835    1.2073    1.2087    0.0014
  15.2294    1.3163    1.3155    0.0008

A =
    0.9997
B =
   -0.0066
C =
   8.9906e-04
D =
   5.8606e-05
```

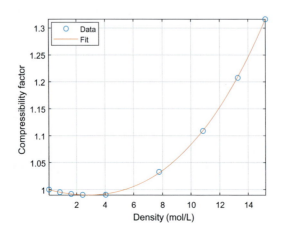

As can be seen, the calculations reveal very good agreement between the measured and the predicted values; the value of coefficient *A* is very close to 1 that coincides with its theoretical value.

5.4.2 Thermal conductivity fitting with the 3rd and 4th degree polynomials

The thermal conductivities *k* of an alloy measured at temperatures *T* closed to absolute zero are: $k = 4.102, 15.7, 23.8, 24, 23.5, 20.2, 11.7$, and $4.2 \, kW/(m\,K)$ and $T = 1, 4, 8, 9, 10, 13, 20$, and $30 \, K$.

Problem: fit the $k(T)$ data with 3rd- and 4th-degree polynomials and calculates discrepancies (i.e., residuals) between the experimental and calculated by the fitting values for each of polynomials. Write the program as user-defined function named ApExample5_2 with the virial coefficients as output parameters and without the input parameters. The function also should generate graph $k(T)$ with the experimental points and two fitted curves. In addition, the program should display the table with the experimental, *T*, and *k* data, and with calculated *k* and the absolute residual values for each of polynomials.

The commands solving this problem are:

```
function [coeff_3,coeff_4]=ApExample5_2
% fits the thermal conductivities vs temperature data with the 3rd degree polynomial
% To run: >>[coeff_3,coeff_4]=ApExample5_2
T=[1 4 8 9 10 13 20 30];                    % vector of the temperatures, K
k=[4.102 15.700 23.800 24.000 23.500 20.200 11.700 4.200];% k-vector, kW/(m.K)
coeff_3=polyfit(T,k,3);                     % 3rd degree polynomial fit
coeff_4=polyfit(T,k,4);                     % 4th degree polynomial fit
kfit3=polyval(coeff_3,T);                   % k by the 3rd degree polynomial
kfit4=polyval(coeff_4,T);                   % k by the 4th degree polynomial
residual3=abs(k-kfit3);                     %residuals: k-kfit3
residual4=abs(k-kfit4);                     %residuals: k-kfit4
Tg=linspace(T(1),T(end));      %vector temperatures for the fitting curve graph
kg_3=polyval(coeff_3,Tg); % therm. cond. values for the 3rd degree pol. curve graph
kg_4=polyval(coeff_4,Tg); % therm. cond. values for the 4th degree pol. curve graph
plot(T,k,'o',Tg,kg_3,'--',Tg,kg_4,'-')      % resulting graph
grid on
xlabel('Temperature, K'),ylabel('Thermal conductivity, W/(m K)')
legend('Data','3rd degree polynomial fit',…
'4th degree polynomial fit','location','best')   % legend is best located on the graph
axis tight
fprintf('   Temp     k    k_fit1 Resid1 k_fit2 Resid2\n') %head of the resulting table
fprintf(' %9.4f %7.4f %7.4f %7.4f %7.4f %7.4f\n',…
[T;k;kfit3;residual3;kfit4;residual4])           % resulting table
```

These commands perform the following actions:

- define the `ApExample5_2` function with the output variables `coeff_3` and `coeff_4`; this function written without input arguments;
- explain the purposes and represent the running command in the two help lines;
- assign the temperature and thermal conductivity values to the `T` and `k` variables;
- fit the $k(T)$ data by the 3rd degree and 4th degree polynomials with two `polyfit` commands and assign defined coefficients to the `coeff_3` and `coeff_4` output variables;
- calculate `kfit3` and `kfit4` by the 3rd and 4th degree polynomials respectively using the `polyvalue` commands with the corresponding fitting coefficients;
- assign, for further plotting, vector of the temperatures `Tg` by the `linspace` command with minimal (as the first element of the `T` vector) and maximal (as the last element of the `T` vector) temperatures; calculate two vectors of the thermal conductivities `kg_3` and `kg_4` for the 3rd and the 4th degrees polynomials respectively using two `polyvalue` commands with the assigned `Tg`;
- plot the $k(T)$ data (circles) and two fitting curves for 3rd degree and 4th degree polynomial fitting (dashed and solid lines), labels the axes and shows the legend in the "better" place;
- display the table caption and six columns with the `T`, `k`, `kfit3` and `residual3` (the 3rd degree fit), and `kfit4` and `residuals4` (the 4th degree fit).

The running command, the resulting table, the fitting coefficients for the 3rd- and 4th-degree polynomials, and generated graph are:

```
>> [coeff_3,coeff_4]=ApExample5_2

    Temp        k       k_fit1    Resid1    k_fit2    Resid2
   1.0000    4.1020    4.2140    0.1120    3.7842    0.3178
   4.0000   15.7000   15.9303    0.2303   16.6354    0.9354
   8.0000   23.8000   22.9491    0.8509   23.1927    0.6073
   9.0000   24.0000   23.4532    0.5468   23.4640    0.5360
  10.0000   23.5000   23.5429    0.0429   23.3273    0.1727
  13.0000   20.2000   21.7534    1.5534   21.0798    0.8798
  20.0000   11.7000   11.0701    0.6299   11.5012    0.1988
  30.0000    4.2000    4.2890    0.0890    4.2175    0.0175

coeff_3 =
    0.0071 -0.4001  5.7560 -1.1490
coeff_4 =
   -0.0002  0.0165 -0.5724  6.8129 -2.4727
```

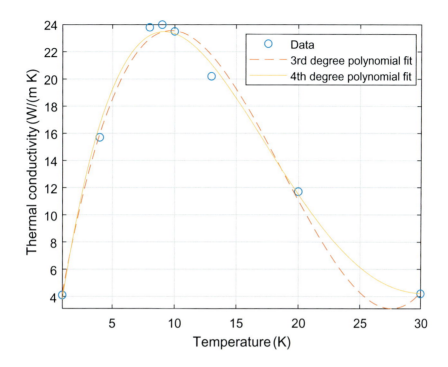

5.4.3 Fitting the stress-strain data by the nonpolynomial equation

For a plastic material, the stress-strain data are: strains $\varepsilon = 0.10, 0.15, 0.33, 0.46, 0.58, 0.68, 0.81$ dimensionless and stresses $\sigma = 41.93, 50.95, 93.22, 115.90, 130.66, 141.49, 154.38$ MPa. The relation that can approximate this data can be represented by the following nonpolynomial equations:

$$\varepsilon = a + b \cdot \ln(\sigma)$$

where a and b are the coefficients that should be defined by fitting.

Problem: Fit the stress-strain data with the above equation using the 1st degree polynomial fit. Write user-defined function with name `ApExample5_2` with the a, b and R-squared (in accordance to the equation in Section 5.1.3) as output- and without input parameters. The function should generate two plots in the same Figure window, the first in semilogarithmic coordinates and second in Cartesian coordinates; in both plots, the data points and the fitting curve should be shown. The table with the stress, strain, fitting strain, and residual values should be displayed.

The commands that solve this problem are:

```
function [a,b,R_squared]=ApExample5_3
% fits the strain-stress non-polynomial relation by the 1st degree polynomial
% To run: >>[a,b,R_squared]=ApExample5_3
close all
sigm=[0.10 0.15 0.33 0.46 0.58 0.68 0.81];        % vector with stress values, ndm
eps=[41.92 50.95 93.22 115.90 130.66 141.49 154.38];% vector with strain values,Mpa
a_b=polyfit(log(sigm),eps,1);      % 1st degree polynomial fit with x as log(sigma)
a=a_b(2);b=a_b(1);                 % assigning a and b via the elements of the a_b vector
eps_fit=a+b*log(sigm);             % strains by the epsilon = a+b*log(sigma)
s1=sum((eps-eps_fit).^2);
s2=sum((eps-mean(eps)).^2);
R_squared=1-s1/s2;                                 % goodness of fit
sigmg=linspace(sigm(1),sigm(end));
epsg=a+b*log(sigmg);                               % strains for plotting
subplot(2,1,1)
semilogx(sigm,eps,'o',sigm,eps_fit)                % resulting semi-logarithmic plot
grid on
xlabel('Stress, MPa'),ylabel('Strain, dimensionless')
legend('Data','fit','location','best')             % legend is best located on the plot
axis tight
subplot(2,1,2)
plot(sigm,eps,'o',sigmg,epsg)                      % resulting plot
grid on
xlabel('Stress, MPa'),ylabel('Strain, dimensionless')
legend('Data','Fit','location','best')             % legend is best located on the plot
axis tight
fprintf(' Stress   Strain    fit   Residual\n')    % caption of the resulting table
fprintf('%9.4f %9.4f %9.4f %9.4f\n',[sigm;eps;eps_fit;(eps-eps_fit)])%resulting table
```

Represented commands:

- define the `ApExample5_3` function with the output variables `a`, `b`, and `R_squared` and without input arguments;
- explain the purposes and represents the running command—in the two help lines;
- assign the stress and strain values to the `sigm` and `eps` variables;
- fit the $\varepsilon(T)$ data by the 3rd degree and 4th degree polynomials with two `polyfit` commands and assign defined coefficients to the `a` and `b` output variables;
- calculate `eps_fit values` by the fitting equation using defined a and b coefficients;
- calculate R-squared and assign this value to the output variable `R_squared`;
- assign, for further plotting, vector of the stresses `sigmg` by the `linspace` command with minimal (as the first element of the `sigm` vector) and maximal (as the last element of the `sigm` vector) temperatures; calculate the `epsg` strain values by the fitting equation with the assigned `sigmg`;
- generate two subplots—in semilogarithmic and Cartesian coordinates—with the $\varepsilon(T)$ data (circles) and fitting curve (solid line) each; each plot has the axes labels and shows the legend that locates in the "`better`" place;

- display the table caption and four columns with the `sigm`, `eps`, `eps_fit` values and with the `(eps-eps_fit)` values (residuals).

The running command, the resulting table, the fitting coefficients, and generated plot are:

```
>> [a,b,R_squared]=ApExample5_3
   Stress      Strain       fit        Residual
   0.1000      41.9200     34.1280      7.7920
   0.1500      50.9500     56.4541     -5.5041
   0.3300      93.2200     99.8688     -6.6488
   0.4600     115.9000    118.1570     -2.2570
   0.5800     130.6600    130.9207     -0.2607
   0.6800     141.4900    139.6793      1.8107
   0.8100     154.3800    149.3120      5.0680
a =
   160.9149
b =
   55.0628
R_squared =
   0.9854
```

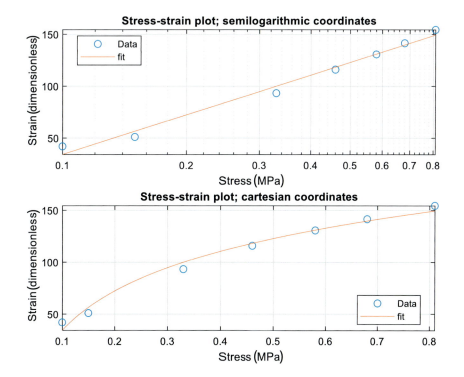

5.4.4 Fitting of the specific heat of a metal with the Basic Fitting interface

The specific heat capacity data of a metal at different temperatures are: C_p=2.159, 2.273, 2.370, 2.478, 2.606, 2.758, and 2.935 cal/(gK) at T=300, 400, 500, 600, 7600, 800, and 900 K.

Problem: Fit this data with the 2nd and 3rd degree polynomials using the Basic Fitting interface and choose the polynomial with the lower "Norm of residual" value. Evaluate the specific heat value at T=200 K by the better polynomial

To use the Basic Fitting interface, first, the $C_p(T)$ graph should be plotted. For this, enter in the Command Window the T and C_p values and then the plot command as follows:

```
>> T=300:100:900;
>> Cp=[2.159 2.273 2.370 2.478 2.606 2.758 2.935];
>> plot(T,Cp,'o')
```

Then, in the opening Figure window, we mark the Basic Fitting line in the popup menu of the Figure window Tools button. In the first panel of the Basic Fitting window, click the "Show the next panel" button → and in the next panel click the → button again. The full Basic Fitting window is opened. Mark now the two following boxes in the "Check to display fit on figure" field of the "Plot fits" panel: "quadratic" and "cubic." Mark also the "Show equations," "Plot residuals," and "Show norm of residuals" boxes. The "Figure window" shows now two following subplots:

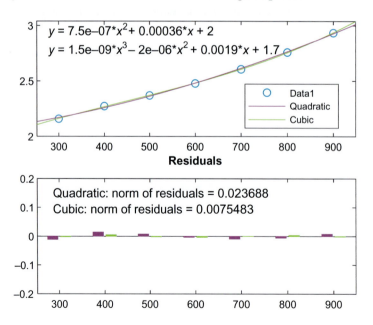

As it can be seen, the 3rd degree polynomial (cubic) has significantly lower norm of residuals than the 2nd degree (quadratic), and therefore, it more accurately fits the

original specific heat data. To evaluate the specific heat value at $T=200$ K, type the value 200 in the "Enter values or …" (shortened) field of the "Find y=f(x)" panel and press the "Evaluate" button, the searching value 2.04 (in cal/(gK)) appears in the "f(x)" field. The "Basic fitting" window with all signed fields/boxes and achieved results are represented below.

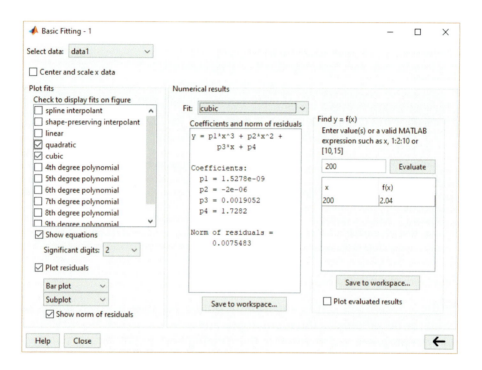

The polynomial, defined coefficients, and norm of residuals are represented in the "Coefficient and norm of residuals" field and in the plots of the Figure window.

5.4.5 Fitting the dynamical viscosity-temperature-pressure data by optimization

The dynamic viscosity μ of an organic compound was measured at various temperatures T and pressures P. The obtained data is $T=323, 323, 333, 333, 343, 343, 353, 353$ K, $P=0.1, 20, 40, 60, 80, 100, 0.1, 100$ MPa, $\mu=0.5, 0.65, 0.7, 0.85, 0.91, 1.06, 0.37, 0.96$ mPa s. The equation that is expected to fit this data is:

$$\mu = aP^c \exp\left(\frac{b}{T}\right)$$

where a, b, and c are the coefficients that should be defined.

Problem: Fit the $\mu(T,P)$ data with the above equation using the fminsearch optimization command. Write user-defined function with name `ApExample5_5` with the *a, b c,* and *R*-squared as output parameters and with the initial *a*0 vector of coefficients as input parameter. Take the initial *a, b,* and *c* values of the *a*0 vector as 1, −10, and 1 respectively. The function should display a table containing the temperature, pressure, and dynamic viscosity data, together with defined viscosities and residuals.

The commands that solve this problem are:

```
function [a,b,c,R_squared]=ApExample5_5(a0)
% fits the viscosities versus temperatures and pressures data
% To run: >>[a,b,c,R_Squared]=ApExample5_5([1 -10 1])
T=[323 323 333 333 343 343 353 353];              % vector of temperatures, K
P=[0.1 20 40 60 80 100 0.1 100];                   %vector of pressures, Mpa
mu=[0.50 0.65 0.70 0.85 0.91 1.06 0.37 0.96];      %vector of viscosities, mPa.s
a_b_c=fminsearch(@(a)myfun(a,T,P,mu),a0);   % obtaining the a, b, and c as vector
a=a_b_c(1);b=a_b_c(2);c=a_b_c(3);  % assigns the a, b, and c to the output variables
mufit=a*P.^c.*exp(b./T);          % calculates the viscosities by the defined equation
Res=mu-mufit;                                     % calculates residuals
R_squared=1-sum(Res.^2)/sum((mu-mean(mu)).^2);     % calculates R-squared
fprintf(' T     P      mu    mu_fit  Residual\n')
fprintf('%7.2f %7.1f %6.2f %6.2f  %7.4f\n',[T;P;mu;mufit;Res])% resulting table
function sse=myfun(a,T,P,mu)         % sub-function for the fminsearch command
sse=sum((mu-a(1)*P.^a(3).*exp(a(2)./T)).^2);   % sum of the squared residuals
```

Represented commands execute the following:

- define the `ApExample5_5` function with the output variables `a, b, c` and `R_squared` and with the a0 variable as the input arguments;
- explain the purposes and represents the running command—in the two first help lines;
- assign the temperature, pressure, and viscosity values to the `T, P,` and `mu` variables;
- fit the $\mu(T,P)$ data with the `fminsearch` optimization command and assign defined coefficients to the `a_b_c` vector; the optimization is performed by searching minimum of the `sse`—squared sums—$\Sigma[\mu-aT^c\exp(b/T)]^2$; sse wrote in the `myfun` subfunction that has the `a, T, P,` and `mu` variables as its input and the `sse` as output;
- assign the defined coefficients to the `a, b,` and `c` output parameters;
- calculate the `mu_fit` values by the fitting equation using the defined *a, b,* and *c* coefficients;
- calculate and assign the residuals to the `res` variable;
- calculate the *R*-squared and assign this value to the output variable `R_squared`;
- display with the `fprintf` command, the table caption and five columns with the `T, P, mu, mufit,` and `res` values.

The running command, the resulting table, the fitting coefficients, and *R*-squared are:

```
>> [a,b,c,R_squared]=ApExample5_5([1 -10 -1])

T          P        mu     mu_fit   Residual
323.00     0.1      0.50   0.37     0.1250
323.00    20.0      0.65   0.70    -0.0501
333.00    40.0      0.70   0.80    -0.1009
333.00    60.0      0.85   0.84     0.0099
343.00    80.0      0.91   0.91    -0.0034
343.00   100.0      1.06   0.94     0.1223
353.00     0.1      0.37   0.44    -0.0655
353.00   100.0      0.96   0.98    -0.0228

a =
   2.8597
b =
  -568.5814
c =
   0.1178
R_squared =
   0.8781
```

The *R*-squared value is lesser than 0.9 that shows a not very good approximation of the viscosities with the approximation equation; another equation probably should be used.

The solution is sensitive to the initial coefficients values, see comments in Section 5.3 on this issue.

5.5 Questions and exercises for self-testing

1. To define fit coefficients with polynomial fit, the following commands should be used: (a) `polyvalue`, (b) `polyfit`, (c) `interp1`, (d) `fittool`. Choose the correct answer.
2. To calculate the polynomial value the following command should be used: (a) `polyfit`, (b) `polyval`, (c) `polyvalm`, (d) `polyvalue`. Choose the correct answer.
3. Fitting by the `polyfit` command brings the coefficient vector p = [0.1234 −0.2543 1.2500]. What does the defined polynomial look like? (a) $1.2500x^2 - 0.2543x + 0.1234$; (b) $0.1234x^2 - 0.2543x + 1.2500$; (c) $1.234x^3 - 0.2543x^2 + 1.2500x$; (c) $1.2500x^3 - 0.2543x^2 + 0.1234x$? Choose the correct answer.
4. To activate the Basic Fitting tool: (a) plot previously (*y, x*)-data points and select the Basic Fitting line in the Tools option of the Figure window menu; (b) click the Curve fitting button in the APPS window of the MATLAB® toolstrip; (c) both answers (a) and (b) are correct; (d) none of the answers is correct.

5. The appearing first panel of the Basic Fitting can be extended to the full-sized window by: (a) choosing the "Separate figure" line in the popup menu of the "Subplot" field of the "Plot fits" panel; (b) clicking the "Show next panel" button of the "Plot fits" panel and then clicking the same button in the "Numerical results" panel; (c) both answers (a) and (b) are correct; (d) marking the "Plot evaluated results" box in the "Find y=f (x)" panel. Chose the correct answer.

6. To fit a data with the equation $y = a_1 + a_2 \cdot \ln(x)$ by the polyfit command, this command should be written as follows: (a) a=polyfit(x,log(y),1); (b) a=polyfit(log(x),y,1); (c) a=polyfit(log(x),log(y),1); (d) a=polyfit(x,y,2). Choose the correct answer.

7. A data were fitted with the equation $y = a_1 \cdot \exp(a_2 x)$ by the polyfit command that outputted coefficients as two-element vector b (i.e., b(1) and b(2)). Which of the following command gives the a_1 and a_2 coefficients: (a) a1=exp(b(2)); a2=b(1); (b) a1=exp(b(1)); a2=b(2); (c) a1=exp(b(1)); a2=log(b(2)); (d) a1=exp(b(2)); a2=log(b(1)). Choose the correct answer.

8. To evaluate the fit, the following criterions can be used: (a) norm of residuals; (b) R-squared; (c) both answers (a) and (b) are correct; (d) none of the answers is correct. Choose the correct answer.

9. Criterion R-squared for 1st degree polynomial is 0.8912 and for 2nd degree polynomial—0.9671. Which polynomial best fits the data? (a) 1st degree polynomial; (b) 2nd degree polynomial; (c) it is difficult to give a defined answer; you need to plot the fitting data together with two polynomial curves and make sure that the best polynomial does not oscillate between the fitting points. Choose the correct answer.

10. The `fminsearch` optimization command is used for fitting the given $Y(x1,x2)$ values by the $Y_{fit} = f(x1,x2)$ equation. In which form the fitting equation should be represented for using with this command? (a) y=Y-Y_{fit}; (b) y=sum(P-Y_{fit}); (c) sse=sum((Y-Y_{fit}).^2); (d) sse=sum((Y-Y_{fit}).^2./n), where n is the number of the Y values. Choose the correct answer.

11. The data on the surface tension σ of a liquid metal as a function of the temperature T are: σ=404.6, 398.5, 392.1, 385.2, 378.1, 370.7 N/m and T=473, 523, 573, 623, 673, 723 K. Write the user-defined function named Problem_5_11 that fits the data by the 2nd degree polynomial: $\sigma = a_2 + a_1 T + a_0 T^2$. The function should have output parameters only: the a_0, a_1, and a_2 that represent the fitted polynomial coefficients and R-squared (see expression in Section 5.1.3) that to be need to be calculated. The fitting coefficient values, the graph representing data and fitting curve, and the resulting table with the original and obtained surface tensions, and the discrepancies (residuals) in the σ-values should be displayed with the `fprintf` function.

12. The rate of the heat transferring from the hot to cold side of the material is called the thermal diffusivity. The obtained values of the thermal diffusivity α of a solid substance as function of the temperature T are: α=32, 25, 19.7, 15.8, 13.1, 11.5 cm^2/s and T=373, 573, 773, 973, 1173, 1373 K. Write the user-defined function named Problem_5_12 that fits the data by the 1st and 2nd degree polynomials ($\sigma = a_1 + a_0 T$ or $\sigma = a_2 + a_1 T + a_0 T^2$). The function has only output parameters: the R-squared

calculated for the both fitting polynomials. The following results should be represented: coefficient values for each of polynomials, the graph representing data and two fitting curves, the resulting table with the original and obtained thermal diffusivities for each polynomial, and the discrepancies (residuals) in the α-values.

13. The force-elongation test data, $F(\Delta l)$, for a rubber are: $F = 4.00, 5.25, 5.52, 6.58, 8.54, 13.88, 18.42, 23.75, 27.67, 31.67, 34.96, 37.45$ N and $\Delta l = 61, 122, 183, 213, 244, 274, 290, 335, 366, 396, 427, 457$ mm. Write the user-defined function named `Problem_5_13` that fits the data by polynomials of the various degrees n. The function should have the degree of polynomial as the input parameter and the fitting coefficients and norm of residuals (see expression in Section 5.1.3) as output parameters. The results should be represented in graph and in table:
    - the graph shows the original points and fitting curve with best-located legend displayed the polynomial degree (to join the characters and the number variables in the one vector use the `['Fit,' num2str(degree) '-degrree pol.']` argument in the `legend` command);
    - the table displays the original data, the fitting rubber elongations, and the residuals. Try running the function with the polynomial degree values are equal to 4, 5, and 6.

14. The dynamic viscosity-temperature data, $\mu(T)$, of an liquid alloy are: $\mu = 2.647, 2.0878, 1.7117, 1.4460, 1.2508, 1.1027, 0.9873$ mPa s and $T = 513, 563, 613, 663, 713, 763, 813$ K. Write the user-defined function named `Problem_5_14` that fits the data by the nonpolynomial equation $\mu = a_0 \cdot \exp(a_1/T)$; the function has the output parameters only—coefficients a_0 and a_1 and norm of residuals. The results should be represented in the graph and in the table:
    - the graph shows the original points, the fitting curve, and the "best"-located legend;
    - the table represents the original data, the fitting alloy viscosities, and the residuals.

    Use the 1st degree polynomial fit with $1/T$ and $\log(\mu)$ as respectively `x` and `y` parameters in the `polyfit` command; to calculate fitting values of the viscosity use the nonpolynomial function with $a_0 = $ `a(2)` and $a_1 = $ `a(1)` where `a` is the output coefficient vector of the `polyfit` command.

15. The yield strength-grain diameter data, $\sigma_y(d)$, for a metal are: $\sigma_y = 0.205, 0.150, 0.135, 0.097, 0.089, 0.080, 0.070, 0.067$ kPa and $d = 5, 9, 16, 25, 40, 62, 85, 110$ µm.

    Write the user-defined function named `Problem_5_15` that fits the data by the nonpolynomial Hall-Petch-type equation $\sigma_y = a_0 + a_1 \cdot d^{-1/2}$. The function has no input parameters and has the coefficients a_0 and a_1 and R-squared as output parameters. The results should be represented in the graph and in the table:
    - the graph shows the original points, the fitting curve, and the "best"-located legend;
    - the table represents the original data, the fitting yield strength, and the percentage of the discrepancies $|(\sigma_y - \sigma_{y\_fit})|/\sigma_y * 100$ where σ_{y_fit} is the yield strength values calculated by the fitting equation.

    Use the 1st degree polynomial fit by the `polyfit` command with $\cdot d^{-1/2}$ and σ_y as the `x` and `y` parameters respectively.

16. The intensities of light I in a transparent solid were measured at different lengths of penetration L of the light. The data are: $I = 3.9, 3.7, 3.5, 3.3, 2.6, 2.2\,\text{W/m}^2$ and $L = 0.19, 0.8, 1.3, 1.8, 4.1, 5.6\,\text{cm}$ respectively; the intensity of the incident beam I_0 was equal to $4.7\,\text{W/m}^2$. Write the user-defined function named `Problem_5_16` that fits the data by the nonpolynomial equation $I = a_0 \cdot I_0 \cdot \exp(-a_1 L)$ where a_1 is the absorption coefficient. The function has no input parameters and has the output parameters—coefficients a_0 and a_1 and R-squared. The results should be represented in the graph and in the table as follows:
- the graph shows the original points, the fitting curve, and the "best"-located legend;
- the table represents the original data, the fitting light intensities, and the residuals.

Use the 1st degree polynomial fit by the `polyfit` command with L and $\log(I)$ as its the x and y parameters respectively; to calculate fitting values of the viscosity use the nonpolynomial function with $a_0 = \exp(a(2))/I_0$ and $a_1 = a(1)$ where a is the output coefficient vector of the `polyfit` command.

17. The Prandtl number Pr represents the dimensionless ratio of momentum diffusivity to thermal diffusivity. The observed Prandtl number-temperature data, $Pr(T)$ of a fluid metal are: $Pr = 0.0116, 0.0111, 0.0108, 0.0105, 0.0104, 0.0104, 0.0105, 0.0107, 0.0110$ and $T = 833, 863, 893, 923, 953, 983, 1013, 1043, 1073\,\text{K}$. Fit this data with the quadratic polynomial using the Basic Fitting tool. Generate with this tool two subplots in the same Figure window: the first—with data, fitting curve, and fitting equation with four significant digits; the second—with residuals as bars and norm of residuals.

Save the results in the MATLAB® workspace.

18. The isobaric heat capacity-temperature data, $C_p(T)$, of an alloy are: $C_p = 183.0, 182.7, 182.4, 182.0, 181.7, 181.4, 181.2, 180.9\,\text{J/(kg K)}$ and $T = 513, 563, 613, 663, 713, 763, 813, 863\,\text{K}$. Fit this data with the linear and quadratic polynomials using the Basic Fitting tool. Generate with this tool two subplots in the same Figure window: the first plot should represent data, two fitting curves, and two fitting equations with coefficients having the four significant digits, and the second—residuals as line curve with two norm of residuals. Which polynomial is more accurate? Save all results in the MATLAB® workspace.

19. The vapor pressure-temperature data, $p_s(T)$, of a liquid alloy are: $p_s = 9.72\text{e-}05$, $0.000535, 0.00223, 0.00754, 0.0214, 0.0530, 0.117, 0.237\,\text{MPa}$ and $T = 513, 563, 613, 663, 713, 763, 813, 863\,\text{K}$. Fit this data with the following equation: $p_s = \exp(a_0/T + a_1)$ using the Basic Fitting tool. For this, present this equation in form $\ln(p_s) = a_0/T + a_1$, transform the p_s and T to the $\ln(p_s)$ and $(1/T)$ data, and generate corresponding plot. Realize fit in the Basic Fitting window and save obtained coefficients in the workspace. Write a script that uses the fitting coefficients to calculate the vapor pressures, and to display table with the original and calculated p_s values.

20. The thermal conductivity λ of a siliciclastic material were measured at different temperatures T and pressures P. The data are: $T = 273, 273, 323, 323, 373, 373, 423,$

423 K, $P = 0.1, 50, 100, 150, 200, 250, 250$ MPa, $\lambda = 1.01, 2.05, 1.99, 2.01, 1.96, 1.98, 1.92$ W/(m K). The equation, which as expected may fit the data, is:

$$\lambda = \frac{1 + \alpha P}{A + BT}$$

where α, A, and B are coefficients.

Write user-defined function with name Problem_5_20 that fits the data to the above equation. The function has the `a0` input parameter for the initial coefficient values and has the output `Alpha`, `A` and `B` variables for fitting coefficients α, A, and B respectively. Fit the $\lambda(T,P)$ data using the `fminsearch` optimization command. Take the following vector of the initial coefficient values: a0 = [1 2 1]. The function should display a table with the original and obtained data, and residuals.

21. Electromagnetic relative permeability μ of a ferrite sample is defined in the inductance L measurements at various temperatures T. The obtained μ, L, T—data are: $T = 298, 308, 318, 328, 338, 348, 358$ K, $L = 42.03\ 41.69, 42.07, 40.44, 35.19, 19.49, 2.028$ mH, $\mu = 849.7, 842.2, 849.9, 817.0, 710.9, 393.7, 41.0$ dimensionless. The equation that as expected can fit these data:

$$\mu = \frac{L}{C(T-B)^\gamma}$$

where C, B, and γ are the coefficients that should be defined in fitting.

Write user-defined function with name Problem_5_21 that fits the data to the above equation. The function has the `a0` input parameter for the initial values of the C, B, and γ coefficients and has the output C, B, Gamma, and R_squared variables for the fitting coefficients C, B, γ and the R-squared respectively. Fit the $\mu\mu(L,T)$ data using the `fminsearch` optimization command. Take the following vector of the initial coefficient values: a0 = [1e-2 1 200]. The function should display a table with the original and obtained data, and residuals.

5.6 Answers to selected questions and exercises

2. (b) `polyval`.
4. (a) plot previously (y, x)-data points and select the Basic Fitting line in the Tools option of the Figure window menu.
6. (b) a=polyfit(log(x),y,1);
8. (c) both answers (a) and (b) are correct.
10. (c) sse=sum((Y-Y$_{fit}$).^2);

12.

```
>> [R_sq_1st_degree,R_sq_2nd_degree]=Problem_5_12
The 1st degree polynomial a0=-0.0203 a1=37.2386
The 2nd degree polynomial a0=48.0155 a1=-0.0495 a2= 0.0000

    T     Sigma   Fit_1    Res_1    Fit_2    Res_2
   373    32.0    29.7     2.3333   31.9     0.1071
   573    25.0    25.6    -0.6067   25.2    -0.1614
   773    19.7    21.5    -1.8467   19.8    -0.0657
   973    15.8    17.5    -1.6867   15.7     0.0943
  1173    13.1     3.4    -0.3267   13.0     0.1186
  1373    11.5     9.4     2.1333   11.6    -0.0929

R_sq_1st_degree =
   0.9452
R_sq_2nd_degree =
   0.99976
```

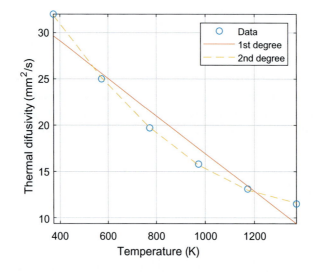

14.

```
>> [a0,a1,Norm_res]=Problem_5_14

    T      mu    Fit    Residual
   513    2.6    2.6    0.0028
   563    2.1    2.1    0.0011
   613    1.7    1.7    0.0022
   663    1.4    1.4    0.0011
   713    1.3    1.3    0.0011
   763    1.1    1.1    0.0001
   813    1.0    1.0    0.0030
```

```
a0 =
  0.18284
a1 =
  1371.1
Norm_res =
  9.4205e-05
```

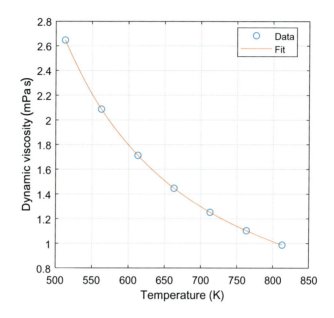

16.

```
>> [a0,a1,R_squared]=Problem_5_16
     L     I     Fit    Residual
     0    3.9    3.9    -0.0265
     1    3.7    3.7     0.0203
     1    3.5    3.5     0.0111
     2    3.3    3.3    -0.0081
     4    2.6    2.6     0.0103
     6    2.2    2.2    -0.0075

a0 =
  0.8525
a1 =
  -0.1064
R_squared =
  0.9993
```

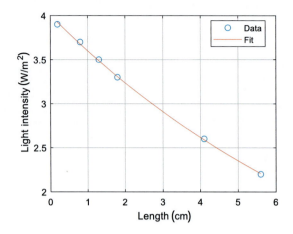

18.
Before opening the Basic Fitting tool:

>> Cp=[183.0 182.7 182.4 182.0 181.7 181.4 181.2 180.9]
>> T=[513 563 613 663 713 763 813 863];
>> plot(T,Cp,'o'),grid on
>> title('Heat capacity as function of temperature')
>> ylabel('Heat capacity, J/(kg.K)')

The Basic Fitting tool with the signed fields solving the problem.

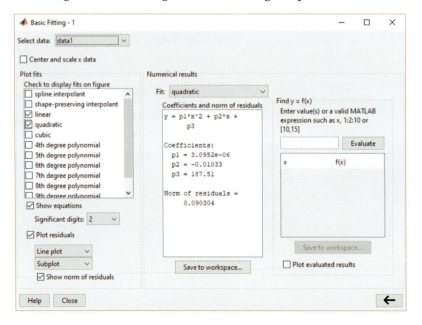

Resulting graph:

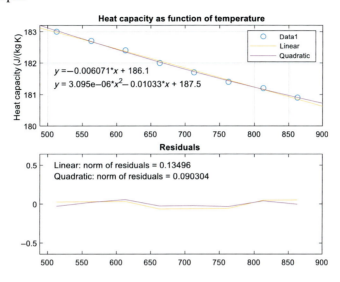

The variables saved in the workspace and its values (were copied and pasted into the Editor page).

```
% fit1 =
%   struct with fields:
%     type: 'polynomial degree 1'
%     coeff: [-0.0061 186.0896]
% normresid1 =
%     0.1350
% resids1 =
%     0.0250
%     0.0286
%     0.0321
%    -0.0643
%    -0.0607
%    -0.0571
%     0.0464
%     0.0500
%---------------------------------
% fit2 =
%   struct with fields:
%     type: 'polynomial degree 2'
%     coeff: [3.0952e-06 -0.0103 187.5141]
% normresid2 =
%   0.0903
% resids2 =
%    -0.0292
%     0.0208
%     0.0554
%    -0.0256
%    -0.0220
%    -0.0339
%     0.0387
%    -0.0042
```

20.

```
>> [al,A,B]=Problem_5_20([1 2 1])
       T          P         L      Fit    Residual
    273.00       0.10      2.01    2.02    0.0084
    273.00      50.00      2.05    2.04    0.0098
    323.00     100.00      1.99    1.99    0.0006
    323.00     150.00      2.01    2.01    0.0004
    373.00     200.00      1.96    1.96    0.0024
    373.00     250.00      1.98    1.98    0.0027
    423.00     250.00      1.92    1.92    0.0026
    423.00       0.10      1.82    1.82    0.0010
al =
  2.1655e-04
A =
  0.3966
B =
  3.6200e-04
```

6

ODE-, PDEPE-solvers, and PDE Modeler tool with applications

Differential equations (DEs) play a prominent role in science and technology in general and in materials science and engineering in particular. They can describe various processes and phenomena in natural sciences taught in colleges and universities and used by the MSE students and specialists. However, actual DEs are often unsolvable analytically. In this case, the sole possibility is a numerical solution. Unfortunately, there is no universal numerical method. Thus, MATLAB® provides a special means, called *solvers*, for solving two groups of differential equations, specifically ordinary, ODE, and partial, PDE. The corresponding groups of solvers are described in brief below, namely the ODE and PDEPE included in the basic MATLAB® software and PDE Modeler tool attached to the PDE toolbox. Sufficient familiarity with both categories of the PDEs is assumed.

6.1 Ordinary differential equations and ODE solver

The single or multiple ordinary differential equations for which the ODE solver is intended are:

$$\frac{dy_1}{dt} = f_1(t, y_1, ..., y_n)$$

$$...$$

$$\frac{dy_n}{dt} = f_n(t, y_1, ..., y_n)$$

where n is the number of first order ODEs, $y_1, ..., y_n$ are the dependent variables while t is the independent that varies between the starting t_s and final t_f values; instead of t, the variable x can be also used. The high-order ODEs must be reduced to the first order by creating a set of first-order equations. For example, for the second order ODE, the first derivative dy_1/dt is denoted as y_2 and than the y_2 is substituted in the original equation. This procedure can be continued in case of the high-order ODEs up to lowering the DE to a set of the first order ODEs.

For example:

- The second order equation

$$\frac{d^2y}{dt^2} + \frac{1}{4}\left(\frac{dy}{dt}\right)^2 y = \cos(t)$$

after denoting y as y_1 and $\frac{dy_1}{dt}$ as y_2 can be transformed to two first-order differential equations:

$$\frac{dy_1}{dt}=y_2, \quad \frac{dy_2}{dt}=-\frac{1}{4}y_2^2 y_1+\cos(t)$$

Now the $y_1(t)$ and $y_2(t)$ values should be defined. The $y_1(t)$ is the solution of the original second order equation while $y_2(t)$ is the derivative of the solution.

- The third order equation:

$$\frac{d^3y}{dt^3}+a\left[\left(\frac{dy}{dt}\right)^2+\frac{y}{t}\right]=\cos(t)$$

should be transformed to a set of three first order equations:

$$\frac{dy_1}{dt}=y_2, \quad \frac{dy_2}{dt}=y_3, \quad \frac{dy_3}{dt}=-a\left[y_2^2+\frac{y_1}{t}\right]+\cos(t)$$

In this case, the $y_1(t)$, $y_2(t)$, and $y_3(t)$ values must be defined. The $y_1(t)$ is solution of the original third order equation while $y_2(t)$ and $y_3(t)$ are, respectively, the first and second derivatives of the solution.

- The fourth order equation:

$$\frac{d^4y}{dt^4}+b\frac{d^2y}{dt^2}+c\left(\frac{dy}{dt}\right)^2 y=\alpha y$$

should be transformed to the set of fourth first order equations:

$$\frac{dy_1}{dt}=y_2, \quad \frac{dy_2}{dt}=y_3, \quad \frac{dy_3}{dt}=y_4, \quad \frac{dy_4}{dt}=-by_3-cy_2^2 y_1+\alpha y$$

For this case, the $y_1(t)$, $y_2(t)$, $y_3(t)$, and $y_4(t)$ values should be defined. The $y_1(t)$ is the solution of the original fourth order equation while $y_2(t)$, $y_3(t)$, and $y_4(t)$ are, respectively, the first, second, and third derivatives of the solution.

Nevertheless, among the practical MSE problems, the first or second order ODEs predominate. Thus, the solution of such equations is discussed further.

There is no universal method for the numerical solution of any of ODEs. Therefore, the MATLAB® ODE solver contains a number of commands realizing different numerical methods for solving actual ODEs.

6.1.1 About numerical methods for solving ODEs

The governing technique used for the solution of the ODE is representing the first-order equation derivative $\frac{dy}{dt}$ as the ratio of the finite differences, i.e.:

$$\frac{dy}{dt}\to\lim_{\Delta t\to 0}\frac{\Delta y}{\Delta t}=\frac{y_{i+1}-y_i}{t_{i+1}-t_i}$$

where Δt and Δy are the argument and function differences and i is the point number in the [a,b]—range of the argument t. This representation is apparently true for very small, but finite (nonzero) distances between the $i+1$ and i-th points.

The procedure for numerical solving the first-order DE, $\frac{dy}{dt} = f(t,y)$ is to give the initial, first-point value y_0 at t_0 and calculate the value of derivative at t_0:

$$\left.\frac{dy}{dt}\right|_0 = f(t_0, y_0).$$

Giving the argument difference Δt, the next function value y_1 can be determined as:

$$y_1 = y_0 + \left[\frac{dy}{dt}\right]_0 \Delta t$$

Now we can calculate derivative at $t_1 = t_0 + \Delta t$:

$$\left.\frac{dy}{dt}\right|_1 = f(t_1, y_1)$$

For the next argument, t_1 we determine y_2:

$$y_2 = y_1 + \left[\frac{dy}{dt}\right]_1 \Delta t$$

Such calculations are repeated for all functions values y_i in the range $t_0 \ldots t_n$ (here n is the number of the final t-point).

This approach, implemented by Euler, is used with various changes, improvements, and complications in advanced numerical methods such as the Runge-Kutta, variable order differences, Rosenbrock and Adams.

The represented scheme describes so-called initial-value problems, IVP, when the DE is solved with any initial value of the function, e.g., $y=0$ at $t=0$ for the $\frac{dy}{dt} = f(t,y)$ equation. The problem when ODEs are solved with two boundary values specified at the opposite range ends is called a boundary-value problem, BVP. The latter problem lies beyond the scope of the book and is not discussed here.

6.1.2 The ode45 and ode15s commands for solving ODEs

The most frequently used commands to the ODE solving have the form:

```
[t,y]=odeN(@function_with_ode,tspan,y0)
```

where

- `odeN` denotes the `ode45` (Runge-Kutta method) or `ode15s` (variable order differences method) commands.
- `@function_with_ode` is the name of the user-defined function/file where the solving differential equations are written. The definition line of the user-defined function should have the form:
  ```
  function dy= function_with_ode (t,y)
  ```

where t, y, and dy are the argument, function, and differential of the function, respectively.

In the subsequent lines, the differential equation/s, presented as first-order PDEs (see Section 6.1), should be written in the form:

```
dy=[right side of the first ODE
right side of the second ODE
...];
```

- tspan—a row vector specifying the integration interval, e.g., the [1 12] vector specifies a *t*-interval with the starting and final *t*-values equal to 1 and 12, respectively. This vector can contain more than two values intended to display the solution at these values, e.g., [1:2:11 12] means that results of solution lays in the *t*-range of 1 … 12 and will be displayed at *t*-values of 1, 3, 5, 7, 9, 11, and 12. The values given into tspan affect the output but not the solution tolerance. The solver command automatically chooses and changes the step to ensure the solution tolerance. The default tolerance is 0.000001 absolute units.
- y0 is a vector of initial conditions, e.g., for the set of two first-order differential equations with initial function values $y_1=0$ and $y_2=4$, y0=[0 4]; this vector can be given also as a column, e.g., y0=[0;4].
- y and t are the column vectors with defined *y*-values at the corresponding *t*-values; in case of the second or higher order ODEs, *y* is the matrix with *n*-columns presenting the obtained *y* values to each of the *n* solved first-order ODEs.

The odeN commands can be used without the t and y output arguments. In this case, the solver presents a solution in the automatically generated plot.

The ODE solver commands can solve so-termed nonstiff and stiff problems. The ODE is referred to as stiff when it contains the terms leading to any singularities, e.g., jumps of the function, holes, ruptures, or other. It results in a divergence of numerical solutions, even at very small steps. In contrast to such equations, nonstiff ODEs are characterized by a stable convergence solution.

Note: The ode45 command is intended to solve nonstiff ODEs and ode15s to the stiff ODEs.

Unfortunately, it is impossible to define a priori if the equation is stiff or not and therefore impossible to determine in advance which ODE command should be used. When we solve the ODE that simulates the real technology or phenomenon, we can prompt it to select the certain ODE solver. For example, substance phase transitions, volatile processes, fast chemical reactions, and explosions are apparently described with the stiff equations. Thus, the ode15s command should be used. Sometimes, a ratio of the maximal to minimal coefficients of the ODE is used as a criterion of the stiffness. If this ratio is larger than 1000, the problem is categorized as stiff. This criterion is an empirical and not always correct. Therefore, if the type of the ODE is not known a priori, it is recommended to first use the ode45 command and then the ode15s.

6.1.2.1 ODE solution steps
1. First of all, the original equation should be lowered to the first order ODE (if it is higher than first order) and presented as a set of the equations having the form:

$$\frac{dy}{dt} = f(t, y), \quad t_s \leq t \leq t_f$$

(t_s and t_f are the starting and final times) with initial condition:

$$y = y0 \text{ at } t = t_s = t0$$

given for the each of the first-order equations.

We demonstrate this and the next steps by an ODE for the so-termed unforced second order system:

$$a\frac{d^2y}{dt^2} + b\frac{dy}{dt} + cy = 0$$

This equation describes many different phenomena, including behavior of materials within some viscoelasticity simulations, RC circuits of the various test apparatuses, behavior of dynamical systems, and other technological and physical processes. The y in this equation is the dependent variable denoting the seeking function at time t (independent variable), and a, b, and c are the constants; e.g. $a=0.5$, $b=2$, and $c=0.7$. The time range is $0 \ldots 14$ with initial values being $y0_1 = 0.1$ and $y0_2 = 0.5$. These values are given as nonunits because the equation and its solution can be used for different actual objects. For a numerical solution with the one of ODE solver commands, the equation should be rewritten as:

$$\frac{d^2y}{dt^2} = -\frac{b}{a}\frac{dy}{dt} - \frac{c}{a}y$$

Now denoting y as y_1 and $\frac{dy_1}{dt}$ as y_2, we obtain the following two ODEs of the first order:

$$\begin{cases} \dfrac{dy_1}{dt} = y_2 \\ \dfrac{d^2y_2}{dt^2} = -\dfrac{b}{a}y_2 - \dfrac{c}{a}y_1 \end{cases}$$

2. In the second step, function files containing the user-defined function with solving equation should be created. The definition line of this function must include input arguments t and y and output argument dy that denotes the left side of the first order ODE/s, $\frac{dy}{dt}$. In the next lines, the ODE/s should be written as vector of right-hand parts in a separate line each or in the same line but divided with the semicolon (;). In our example, the function containing the differential equations is:

```
function dy=myODE(t,y)
a=0.5; b=2; c= 0.7;
dy=[y(2);-b/a*y(2)-c/a*y(1)];
```

This function should be saved in an m-file with the myODE name.

Note: If the t-argument is absent in the right part of the differential equation, instead of this argument, the tilde (~) character can be written. The function definition line may appear as follows: `function dy=myfODE(~,y)..`

3. In the final step, the ODE solver command should be chosen and implemented. For our example, when specific recommendations about the stiffness of the equation to be solved are lacking, the `ode45` solver should be selected. In the Command Window, the following commands should be typed and entered:

```
>> [t,y]=ode45(@myODE,[0 :2:14],[.1 ;.5])   % t=0…14, y0_1=0.1,y0_2=0.5

t =
    0
    2
    4
    6
    8
   10
   12
   14
y =
   0.1000   0.5000
   0.1229  -0.0472
   0.0567  -0.0220
   0.0261  -0.0101
   0.0120  -0.0047
   0.0055  -0.0021
   0.0026  -0.0010
   0.0012  -0.0005
```

The process starts with initial values of $y_1=0.1$ and $y_2=0.5$ at $t=0$; the starting and final times of the process are 0 and 14, respectively, with the time step being 2 (to shorten the output). Thus, the vector of the time span was inputted as [0:2:14].

The given commands solve the equation and display the one-column vector of the t values and two-column matrix with the resulting y values. To plot this solution, the `plot` command should be used, but the achieved eight t,y-points are not sufficient to generate a smoothed plot. To produce the plot, the `t_range` should be inputted with the starting and final values only. In this case, the default `t`-steps values will be automatically chosen. The number of points will be sufficient to generate a smoothed $y(t)$ plot. The commands that realize this purpose are:

```
>> [t,y]=ode45(@myODE,[0 14],[.1 ;.5]));      % t=0…14 with default step
>> plot(t,y(:,1),t,y(:,2))
>> xlabel('Time'),ylabel('Y values')
>> title('Solution of the differential equation for the first order system')
>> legend('y','y''')
>> grid on
```

The generated plot is shown in Fig. 6.1.

In the described solution, some commands representing the solving DEs were written and saved in the myODE function file while other of commands—the ODE45, the plot, and plot formatting—were written in the Command Window. It is more reasonable to create a single program file that includes both these parts. For this purpose, a user-defined function file that includes a subfunction with the solving ODE/s should be written. For the discussed example, this file—named SecondOrderODE—reads as follows:

```
function t_y=SecondOrderODE(ts,tf,y0)
% Solution of the second order ODE
% to run: >> t_y=SecondOrderODE(0,14,[.1;.5])
t_range=[ts,tf];
[t,y]=ode45(@myODE,t_range,y0);
plot(t,y)                    % plots two curves correspondingly to two columns of y
xlabel('Time'),ylabel('Y values')
title('Solution of the viscoelastic-type second-order ODE')
legend('y(t)','y(t)''')
grid on
t_y=[t y]; %to output results as three-column table: one t-column and two y-columns
function dy=myODE(~,y)       % sub-function with the ODEs to be solved
a=0.5; b=2; c= 0.7;
dy=[y(2);-b/a*y(2)-c/a*y(1)];
```

The following command for running this program should be entered in the Command Window:

```
>>t_y=SecondOrderODE(0,14,[.1;.5])
t_y =
        0   0.1000   0.5000
   0.0100   0.1049   0.4789
   0.0201   0.1096   0.4585
   ...
```

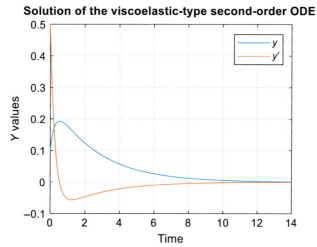

FIG. 6.1 The solution of the second order ODE.

When the command is entered, a three-column table is displayed—one column with *t* and two columns with *y* and *y′* values. Each column has 97 rows (shortened here to save the space). Additionally, the graph (Fig. 6.1) with *y* and *y′* as function of time is plotted.

6.1.2.2 Extended command forms of the ODE solver

When differential equations have parameters (e.g., in the discussed example, the *a*, *b*, and *c* variables are parameters), the ODE solver commands should be written in a more complicated form to transmit the parameters to the function containing the ODEs:

```
[t,y]=odeN(@function_with_ode,tspan,y0,[ ],
           param_name1,param_name2,...)
```

where [ ] denotes an empty vector marking the place for the so-termed options used to control the integration process.[a] In most cases, the default options values are used, which yield satisfactory solutions. Thus, we do not use these options. `param_name1, param_name2, ...` are the names of the arguments that we intend to use for transmitting their values into the `ode_function`.

If the parameters are named in the `odeN` solver, they should also be written into the function containing the ODEs.

For example, the `SecondOrderODE` file should be modified for introducing the *a*, *b*, and *c* coefficients as arbitrary parameters:

```
function t_y=SecondOrderODE(a,b,c,ts,tf,y0)
% Solution of the second order ODE, transfer a, b, and c to the myODE subfunction
% to run: >> t_y=SecondOrderODE(.5,2,.7,0,14,[.1,.5])
close all
t_range=[ts,tf];
[t,y]=ode45(@myODE,t_range,y0,[],a,b,c);
plot(t,y)                % plots two curves correspondingly to two columns of y
xlabel('Time'),ylabel('Y values')
title('Solution of the viscoelastic-type second-order ODE')
legend('y(t)','y(t)''')
grid on
t_y=[t y];% to output results as three-column table: one t-column and two y-columns
function dy=myODE(~,y,a,b,c)        % sub-function with the ODEs to be solved
dy=[y(2);-b/a*y(2)-c/a*y(1)];
```

In the Command Window, the following command should be typed and entered:

```
>> t_y=SecondOrderODE(.5,2,.7,0,14,[0.1;0.5])
```

The results are identical to those discussed in Section 6.1.2.1.

The advantage of the form with parameters is its greater universality, e.g., the `SecondOrderODE` function with parameters can be used for any *a*, *b*, and *c* without introducing its values into the myODE subfunction.

---

[a] Enter the `help odeset` command to review the possible options.

Another, possible simpler form of the commands of ODE solver is:

`[t,y]=odeN(@(t,y)myODE,tspan,y0)`

where `myODE` is the vector containing the right part/s of the ODE/s to be solved. The other arguments are the same as the previously discussed forms. This vector can be written separately with an assigned variable name, i.e., f=@(t,y)myODE. This name should be written within the ode-command, i.e., [t,y]=odeN(f,tspan,y0).

When this form is used, it is not necessary to create the subfunction with ODE/s. For our example, the `SecondOrderODE` function should be rewritten as:

```
function t_y=SecondOrderODE(a,b,c,ts,tf,y0)
% Solution of the second order ODE without the myODE sub-function
% to run: >> t_y=SecondOrderODE(.5,2,.7,0,14,[.1;.5])
close all
t_range=[ts,tf];
dy=@(t,y)[y(2);-b/a*y(2)-c/a*y(1)];            % the ODEs to be solved
[t,y]=ode45(dy,t_range,y0);
plot(t,y)                 % plots two curves correspondingly to two columns of y
xlabel('Time'),ylabel('Y values')
title('Solution of the viscoelastic-type second-order ODE')
legend('y(t)','y(t)''')
grid on
t_y=[t y];% to output results as three-column table: one t-column and two y-columns
```

This form allows us to transmit parameter values to the ODEs without writing them in the `ode45` (or `ode15s`) and in the subfunction with equations. Additionally, this form can be used directly into the Command Window, e.g., for our example:

```
>> [t,y]=ode45(@(t,y) [y(2);-4*y(2)-1.4*y(1)],[0 14],[0.1;0.5]); [t y]
% for output
ans =
         0    0.1000    0.5000
    0.0100    0.1049    0.4789
    0.0201    0.1096    0.4585
    ... (shortened)
```

Application of the represented form is preferable when a single or two short differential equations must be solved.

6.1.3 Supplementary commands of the ODE solver

In addition to the commands studied above, the solver has additional commands that can be used to solve the ODEs; some of these are listed in Table 6.1. The table represents the name of the available ODE solver commands, the numerical methods they use with the type of equations to be solved and examples. The form of the represented commands is identical to those described previously and therefore not specified in the table.

Table 6.1 The supplementary commands for solving ODEs.

Command name	Method, type of the solving equation, and when to use	Example

ode23	Explicit Runge-Kutta method. Nonstiff and moderately stiff problems. Quicker, but less precise than ode45	ODE: y'=2t, with y0=0, ts=0, tf=5.
		`>> [t,y] = ode23(@(t,y) 2*t, [0 5], 0);`
		`>> plot(t,y,'o-')`

ode113	Adams' method. Nonstiff differential equations.
	For problems with stringent error tolerances or for solving computationally intensive problems.
		ODE: the same as for the ode23 example.
		`>> [t,y] = ode113(@(t,y) 2*t, [0 5], 0);`
		`>>plot(t,y,'o-')`

ode23s	Rosenbrock's method. Stiff differential equations. When ode15s is slow.	ODE: y'=-10t, with y0=1, ts=0, tf=2.
		`>> [t,y] = ode23s(@(t,y) -10*t, [0 2], 1);`
		`>> plot(t,y,'o-')`

Chapter 6 • ODE-, PDEPE-solvers, and PDE Modeler tool

Table 6.1 The supplementary commands for solving ODEs—cont'd

Command name	Method, type of the solving equation, and when to use	Example
ode23t	Trapezoidal rule. Moderately stiff differential- and differential algebraic equations, DAEs.	ODE: the same as for the ode23s example. `>> [t,y] = ode23t(@(t,y) -10*t, [0 2], 1);`
ode23tb	Trapezoidal rule/second order Backward differentiation formula, TR/BDF2. Stiff differential equations; sometimes more effective than ode15s.	ODE: the same as for the ode23s example. `>> [t,y] = ode23tb(@(t,y) -10*t,[0 2], 1);`

Based on Table 7.1 in Burstein, L. (2015). *Matlab in Quality Assurance Sciences.* Cambridge: Elsevier-WP.

6.2 Solving partial differential equations with PDE solver

Partial differential equations (PDE) are relevant and applied in many MSE disciplines. In view of this, a short introduction to the MATLAB® PDE solver is provided here. This solver is designed to solve spatially one-dimensional partial differential equations that correspond to the following standard form:

$$c\left(x, t, u, \frac{\partial u}{\partial t}\right) \frac{\partial u}{\partial t} = x^{-m} \frac{\partial}{\partial x}\left(x^m f\left(x, t, u, \frac{\partial u}{\partial x}\right)\right) + s\left(x, t, u, \frac{\partial u}{\partial x}\right)$$

where time t is defined in the range t_0 (initial) ... t_f (final) and the coordinate x between $x = a$ and $x = b$; m can be 0, 1 or 2 corresponding to slab, cylindrical, or spherical symmetry, respectively; $f\left(x, t, u, \frac{\partial u}{\partial x}\right)$ is called a flux term while $s\left(x, t, u, \frac{\partial u}{\partial x}\right)$—a source term. General form of this equation solution is u as function of time and coordinate—$u(x,t)$.

The solver is suitable for elliptical or parabolic types of the PDEs. The following are some examples of PDEs and their compliance to the above equation:

Equation	Form expected by PDE	m	c	f	s
$\frac{\partial T}{\partial t} = \frac{k}{\rho c_p} \frac{\partial^2 T}{\partial x^2} + \frac{q}{\rho c_p}$	$\frac{\partial u}{\partial t} = \frac{\partial}{\partial x}\left(\frac{k}{\rho c_p} \frac{\partial u}{\partial x}\right) + \frac{q}{\rho c_p}$	0	1	$\frac{k}{\rho c_p} \frac{\partial T}{\partial x}$	$\frac{q}{\rho c_p}$
$\frac{\partial \varphi}{\partial t} = D \frac{\partial^2 \varphi}{\partial x^2}$	$\frac{\partial \varphi}{\partial t} = \frac{\partial \varphi}{\partial x} D \frac{\partial^2 \varphi}{\partial x}$	0	1	$D \frac{\partial \varphi}{\partial x}$	0
$\pi^2 \frac{\partial u}{\partial t} = \frac{\partial^2 u}{\partial x^2}$	$\pi^2 \frac{\partial u}{\partial t} = \frac{\partial u}{\partial x}\left(\frac{\partial u}{\partial x}\right)$	0	π^2	$\frac{\partial u}{\partial x}$	0
$\frac{\partial u}{\partial t} = a \frac{\partial^2 u}{\partial x^2} - Fu$	$\frac{\partial u}{\partial t} = \frac{\partial u}{\partial x}\left(a \frac{\partial u}{\partial x}\right) + (-Fu)$	0	1	$a \frac{\partial u}{\partial x}$	$-Fu$
$\frac{\partial u}{\partial t} = \frac{5}{r^2} \frac{\partial u}{\partial r}\left(r^2 \frac{\partial u}{\partial r}\right)$	$\frac{\partial u}{\partial t} = x^{-2} \frac{\partial u}{\partial x}\left(x^2 5 \frac{\partial u}{\partial x}\right)$	2	1	$5 \frac{\partial u}{\partial x}$	0

6.2.1 About numerical methods for solving PDEs

The solution of PDEs by the finite difference method is similar to that described for the ODE and is done by replacing derivatives by the differences. Nevertheless, in the case of a PDE, apart from spatial differences, the time-dependent differences also appear.

We demonstrate this idea by the dimensionless diffusion-type equation $\frac{\partial u}{\partial t} = D \frac{\partial^2 u}{\partial x^2}$, where D is the given constant. Dividing equidistantly the spatial and time intervals into $N+1$ (from 0 to N) and $M+1$ (from 0 to M) points, respectively, this equation can be written as:

$$\frac{u_i^{k+1} - u_i^k}{\Delta t} = D \frac{u_{i+1}^k - 2u_i^k + u_{i-1}^k}{\Delta x^2}$$

where i and k correspond to the current spatial and time point, respectively, while Δx and Δt refer to the x and t differences.

Giving all the u-values at starting time t_s ($k = 0$) and at the boundary, $i = 0$, we can calculate u at time $k = 1$ for the $i = 1$-point:

$$u_1^1 = u_1^0 + D \frac{u_2^0 - 2u_1^0 + u_0^0}{\Delta x^2} \Delta t$$

After this, for $i = 2$:

$$u_2^1 = u_2^0 + D \frac{u_3^0 - 2u_2^0 + u_1^0}{\Delta x^2} \Delta t$$

And so on till the $i = N$:

$$u_N^1 = u_N^0 + D \frac{u_{N+1}^0 - 2u_N^0 + u_{N-1}^0}{\Delta x^2} \Delta t$$

At this point, the u value at the boundary $i=N+1$ is used, which should be given. When all u-values for $k=1$ time point are defined, the next time point $k=2$ can be calculated in the same way.

The discussed scheme, with some improvements and complications, applies to all finite differences methods used to solve the PDE. These schemes and another complex method, called finite elements, are studied in courses on numerical methods. Therefore, their explanations are beyond the scope of this book.

6.2.2 The `pdepe` command for solving one-dimensional PDEs

The command for solving the single set of spatially one-dimensional equations is named `pdepe` and has the following form:

```
sol=pdepe(m,@pde_function,@initial_cond,@bondary_cond,x_mesh,
tspan)
```

where

- `m` is equal to 0, 1, or 2 as above;
- `pde_function` is the name of the function where the solving differential equation must be written; its definition line reads:

```
function [c,f,s]=pdefun(x,t,u,DuDx)
```

the input and output parameters x, t, u, DuDx, c, f, and s being the same x, t, u, $\frac{\partial u}{\partial x}$, c, f, and s as in the standard form expression; note that the derivative $\frac{\partial u}{\partial x}$ of the standard expression is written as DuDx in the pdefun. In the case of two or more PDEs, the c, f, s, and q terms in the standard equation should be written as column vectors. In this case, element-by-element operations should be applied to these vectors.

- `initial_cond` is the name of the function with the initial conditions of the equation being solved:

$$u(x, t_0) = u_0(x)$$

The definition line of this MATLAB® function reads:

```
function u0=initial_cond(x)
```

u0 being the vector $u(x)$ value at $t=0$;

- `boundary_cond` is the name of the function with the boundary conditions

that, for each of the both $x=a$ and $x=b$ (a and b—x-interval ends), boundaries have the following general form:

$$p(x, t, u) + q(x, t)f\left(x, t, u, \frac{\partial u}{\partial x}\right) = 0$$

Note, here the same f as in the standard form equation.

The definition line of the boundary_cond function reads:

`function [pa,qa,pb,qb]=boundary_cond(xa,ua,xb,ub,t)`

xa=a and xb=b being the boundary points, ua and ub are the u values at these points; pa, qa, pb, qb are the same as in the standard form and represent the p and q values at a and b; the way to represent these parameters in real cases is explained later.

- x_mesh is a vector of the x-coordinates at which the solution is sought for each time value contained in the tspan; xmesh values must be written in ascending order from a to b;
- tspan is a vector of the time points; the u-values are searched along all x-points; the t values must be written in ascending order;
- The output argument sol contains defined u-values represented in a three-dimensional array comprising k two-dimensional arrays with i rows and j columns each. Elements in the three-dimensional arrays are numbered analogously to two dimensional (see Section 2.2). For example, sol(1,3,2) denotes a term located in the first row of the third column of the second array (termed sometimes as matrix page) while sol(:,3,1) denotes in the first array (page) all rows of the third column. In the pdepe solver, the sol(i,j,k) denotes defined u_k (u-values for k-th PDE, in case of more than one PDE) at the t_i time- and x_j coordinate points, e.g., sol(:,:,1) is an array whose rows contain the u values defined for all x-coordinates at the certain given time (each row). For a single PDE, there is one array only ($k=1$) for a set of two PDEs—two arrays ($k=1$ and 2, see Section 6.4.2.2) and so on.

More detailed information is obtainable by typing >>doc pdepe or by using the MATLAB® Help.

6.2.3 The steps for solving a PDE with the pdepe command

For example, consider the simple diffusion equation for temporal and spatial changes of a substance concentration u:

$$\frac{\partial u}{\partial t} = D\frac{\partial^2 u}{\partial x^2}$$

where t and x are the time and coordinate, respectively, while D is the diffusion coefficient. All variables in this equation are given in dimensionless units; x ranges from 0 to 1 and t from 0 to 0.4.

This equation is taken with initial and boundary conditions as below.

The initial conditions:

$$u(x,0) = \begin{cases} 1, & 0.45 \leq x \leq 0.55 \\ 0, & \text{elsewhere} \end{cases}$$

The boundary conditions:

$$u(0,t) = u_a = 0, \quad u(1,t) = u_b = 0.$$

The dimensionless diffusion coefficient D is taken as 0.3.

The solution of this single PDE goes through the following steps:

1. First, the equation must be presented in the standard form.

The diffusion equation rewritten to the standard form reads:

$$\frac{\partial u}{\partial t} = \frac{\partial}{\partial x}\left(D\frac{\partial u}{\partial x}\right)$$

Comparing this and the standard equations, we obtain their identity when:

$$m = 0,$$

$$c\left(x, t, u, \frac{\partial u}{\partial t}\right) = 1,$$

$$f\left(x, t, u, \frac{\partial u}{\partial x}\right) = D\frac{\partial u}{\partial x},$$

$$s\left(x, t, u, \frac{\partial u}{\partial x}\right) = 0$$

2. The initial and boundary conditions must be presented in the standard form.

Accordingly, the initial condition:

$$u_0 = \begin{cases} 1, & 0.45 \leq x \leq 0.55 \\ 0, & \text{elsewhere} \end{cases}$$

and boundary conditions at the a and b boundaries, $x = x_a = 0$ and $x = x_b = 1$ are:

$$u_a + 0 \cdot D\frac{\partial u}{\partial x} = 0 \qquad u_b + 0 \cdot D\frac{\partial u}{\partial x} = 0$$

To obtain the latter two equations, the following conventions were used $(0, t, u) = u_a$, $p(1, t, u) = u_b$, $q(0, t) = q_b = 0$, $q(1, t) = q_b = 0$, and $f\left(x, t, u, \frac{\partial u}{\partial x}\right) = D\frac{\partial u}{\partial x}$. Note: in the latter convention, the f function is the same as in the standard PDE.

Generally, examples of the boundary condition recording required by the `pdepe` command can be rewritten to match the standard form $p + qf = 0$. Some real possible boundary conditions and `pdepe`-required p and q values are:

Boundary Conditions	In $p+qf=0$ form	Required p and q
$u=0$	$u=0$	$p=u,\ q=0$
$u=3$	$u-3=0$	$p=u-3,\ q=0$
$D\frac{\partial u}{\partial x}=0$	$D\frac{\partial u}{\partial x}=0$	$p=0,\ q=1$
$D\frac{\partial u}{\partial x}=5$	$D\frac{\partial u}{\partial x}-5=0$	$p=-5,\ q=1$
$D\frac{\partial u}{\partial x}=4-u$	$D\frac{\partial u}{\partial x}-4+u=0$	$p=-4+u,\ q=1$
$g \cdot \frac{\partial u}{\partial x}=h$	$\frac{g}{D} \cdot D\frac{\partial u}{\partial x}-h=0$	$p=-h,\ q=g/D$

The *f* function in all presented cases is the same as in the standard PDE. The *g* and *h* in the last line equations are constants or functions of the *x* and *t*. If the boundary is $\frac{\partial u}{\partial x} = 0$, it is equivalent to the $D\frac{\partial u}{\partial x} = 0$ (after multiplying of both parts of the equation by *D*). The represented conditions can be applied at the *a* and *b* boundaries.

In case of multiple PDEs, the *c*, *f*, and *s* parameters and u0, *p*, and *q* parameters of the initial and boundary conditions, respectively, are column vectors (see Section 6.4.2.2).

3. At this stage, the user-defined function with `pdepe` command and with the three subfunctions containing the solving PDE/s, initial and boundary conditions should be written. To solve the equation in our example, the program named `mydfsn` is:

```
function mydfsn(m,n_x,n_t)
% To run: >> mydfsn(0,100,100)
close all
xmesh=linspace(0,1, n_x);
tspan= linspace(0,0.4, n_t);
u=pdepe(m,@mypde,@i_c,@b_c,xmesh,tspan);
[X,T]=meshgrid(xmesh,tspan);
subplot(1,2,1)
mesh(X,T,u)                                      % mesh plot with the X - T domain
xlabel('Coordinate'),ylabel('Time'), zlabel('Concentration')
title('1D diffusion with the pdepe')
subplot(1,2,2)
c=contour(X,u,T,[0.001 0.005 0.05]);    % contour plot with 3 iso-time lines
clabel(c)                                        % labels for the iso-lines in the contour plot
xlabel('Coordinate'), ylabel('Concentration')
title('Concentrations, iso-time lines')
grid
function [c,f,s]=mypde(x,t,u,DuDx)               % PDE for solution
c=1;
D=0.1;
f= D*DuDx;
s=0;
function u0=i_c (x)                              % initial condition
if x>=0.45&x<=0.55
   u0=1;
else
   u0=0;
end
function [pa,qa,pb,qb]=b_c(xa,ua,xb,ub,t)        % boundary conditions
pa=ua;pb=ub;
qa=0;qb=0;
```

The `mydfsn` function is written without output arguments. Its input arguments are: *m*—as above, n_x and n_t are the serial numbers of the operative *x*- and *t* points of the solution. The designed function contains three subfunctions: `mypde` comprising the diffusion equation, `i_c` defining the initial condition and `b_c` defining the boundary conditions. The x_mesh and tspan vectors are created with two `linspace` commands in which the n_x and n_t must be inputted with the `mydfsn` run. The numerical results are stored in the u array, which in our case is two-dimensional, and used in the subsequent graphical commands. The `mesh` command is used to generate the 3D mesh subplot *u*(*x*,*t*)

(Fig. 6.2A) while the `contour` and `clabel` commands—to generate the 2D subplot $u(x)$ showing the three labeled isotime lines (0.001, 0.03, and 0.3) defined in the `contour` command as the vector of the selected t values (Fig. 6.2B).

After running this function in the Command Window, the following two plots are generated:

```
>> mydfsn(0,100,100)
```

6.3 Partial differential equations with the PDE toolbox interface

While the `pdepe` command solves one-dimensional PDEs, a special tool called PDE Modeler (updated PDE Tool), a part of the Partial Differential Equations Toolbox, is designed to solve two-dimensional PDEs. The method that the modeler uses for PDE solution is called finite elements, FEM. This method is more complicated than the finite difference method but more suitable for the real object geometry and for the inclusion of heterogeneous material properties. The body shape in the x,y plane, referred to as the domain, is divided on small triangles. For each of them, the trial second order polynomial solution is assumed with polynomial coefficients determined from residuals between the trial solution and true solution, which is the DE solution with the polynomial. The residuals are searched so to be zero for each triangle while coefficients are determined from the set of the residual equations. This is a simplified Scheme. A more detailed explanation is beyond the scope of this book.

The PDE Modeler represents an interface for solving elliptic, parabolic, hyperbolic, and eigenvalue types of scalar PDEs with Dirichlet or Neumann boundary conditions. The general standard forms of the PDE-types, boundary conditions, equation name and examples of PDEs with its coefficients correspondence to the standard form coefficients are given below:

Standard form	Name	Example
$m\frac{\partial^2 u}{\partial t_2} + d\frac{\partial u}{\partial t} - \nabla(c\nabla u) + au = f$	Elliptic ($m=0$, $d=0$), parabolic ($m=0$), hyperbolic ($d=0$) PDEs.	$\frac{\partial T}{\partial t} = \nabla(k\nabla T)$, parabolic PDE; $u=T$, $m=0$, $a=0$, $f=0$, $c=k$; $d=1$
$-\nabla(c\nabla u) + au = \lambda du$ or $-\nabla(c\nabla u) + au = \lambda^2 mu$	Eigenvalue PDE.	$-\nabla(\nabla u) = \lambda u$, $c=1$, $a=0$, $d=0$, $m=0$
$hu = r$	Dirichlet boundary: the u value is given.	The $T=0$ boundary: $h=1$, $r=0$.
$\vec{n}(c\nabla u) + qu = g$	Neumann boundary: the du/dx, du/dy and/or u values are given.	The $\frac{dT}{dx}=0$ boundary: $c=1$, $q=0$, $g=0$.

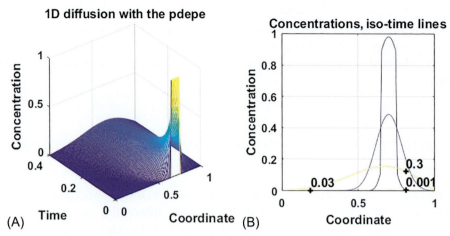

FIG. 6.2 Resulting graphs of the diffusion equation solution with the pdepe command: Concentration-coordinate-time 3D plot (A) and 2D contour plot with three isotime lines (B).

In these equations, $\nabla = \frac{\partial}{\partial x} + \frac{\partial}{\partial y}$—operator nabla; m, d, c, f, a, λ, h, r, q, and g are the coefficients that can be constant or varied with x, y, t, and u; $\vec{n}$ denotes the outward unit normal vector used to indicate the normal direction of the du/dx and du/dy in respect to the surface boundary. The sense of these notations will be further clarified when the actual PDEs are solved with the PDE Modeler.

To open the PDE modeler interface, enter the `pdeModeler` command in the Command Window:

```
>> pdeModeler
```

The window with the PFE Modeler interface appears—Fig. 6.3. The toolstrip of this tool contains the menu, buttons line with frequently used options and the field line to insert the formula. The big empty field with x,y coordinate axes and ticks where the 2D geometry of the actual body for which the PDEs to be solved is to be drawn. The menu contains the File, Edit, Options, Draw, Boundary, PDE, Mesh, Solve, Plot, Window, and Help buttons, each of which has a popup menu with list of additional options. We describe all needed options below together with the PDE solution steps.

6.3.1 Solution steps in the PDE Modeler

To describe the steps of a solution with the PDE Modeler tool, we use the example of a rectangular plate of a solid material with length 3 and width 1 dimensionless units and solve the 2D stationary heat equation:

$$\nabla(k\nabla T) = 0$$

with temperatures $T=1$ (dimensionless) along the two opposite horizontal boundaries of the plate and $T=0$ along the two vertical plate boundaries. The k denotes the

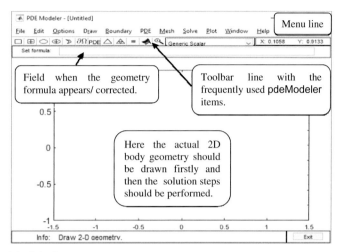

FIG. 6.3 The PDE Modeler interface.

dimensionless material-dependent heat transfer coefficient, which we assume equal to 1 for our plate material.

Step 1. First, the Application mode should be chosen by selecting the appropriate application model in the popup menu of the Application line of the Option button of the main menu—Fig. 6.4.

As can be seen, the Heat Transfer option can be chosen. Nonetheless, for greater generality, the Generic Scalar options (default) are chosen for the further solution.

Step 2. At this stage, the 2D body geometry should be drawn. To do this, the Grid and Snap options (see Fig. 6.4) must be clicked, and the limits of the axes must be settled correspondingly to the sizes of the body being studied. After selecting the Axes Limits or Grid Spacing option, small panels appears with fields requiring the appropriate values. For our plate dimensions, the x-axis limits -0.5 and 3.5 and y-axis limits -0.5 and 1.5 were entered—Fig. 6.5A. The default grid spacing was selected—Fig. 6.5B. The selected limits and grid lines appear in the graph.

The drawing is done by combining shapes, which can be selected from the popup menu of the Draw menu button—see Fig. 6.6.

The same shapes—rectangle, square, ellipse, circle, or polygon—can be selected with a click on the appropriate button of the frequently used button line. To draw our plate, click the rectangle button, press then the mouse left button at point (0,0) point and drag the mouse pointer with the pressed button to the (3,1) grid point; after releasing the mouse button, the rectangle appears—Fig. 6.7.

Instructions for drawing bodies with more complicated geometries using the rectangle, ellipse, or polygon options are presented in Section 6.3.4.

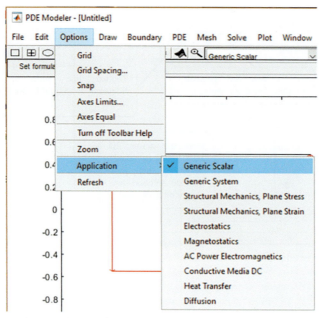

FIG. 6.4 First step in PDE solution; selecting the application mode.

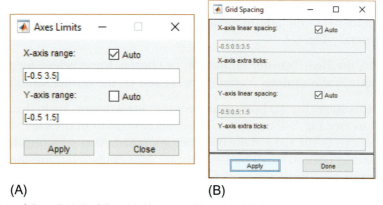

(A)   (B)

FIG. 6.5 The view of the Axis Limits (A) and Grid Spacing (B) panels, filled in as for the equation being solved.

Step 3. In this step, the boundary conditions must be specified. To do this, the Specify Boundary Conditions line should be selected in the popup menu of the Boundary button—Fig. 6.8.

The more suitable way to introduce boundary condition is to click the boundary line on the drawn body. The corresponding type of the boundary conditions should be marked with the necessary boundary equation parameters specified in the appropriate fields. The default boundary condition is the Dirichlet condition.

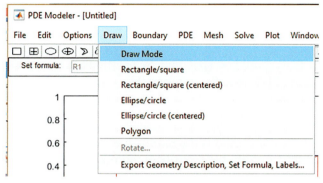

FIG. 6.6 Second step of the PDE solution; selecting the shape of a body.

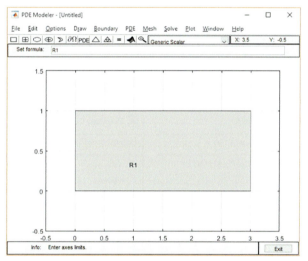

FIG. 6.7 The PDE Modeler window with the rectangle plate drawing.

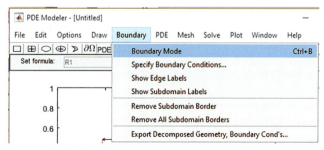

FIG. 6.8 Third step of the PDE solution; introducing the boundary conditions.

226 A MATLAB® Primer for Technical Programming in Materials Science and Engineering

For our problem, for each horizontal side of the rectangle, the following coefficients should be entered: $h=1$ and $r=1$—Fig. 6.9A and $h=1$ and $r=0$ (default) for each of vertical side—Fig. 6.9B.

Step 4. Now the PDE should be specified. Select the PDE Mode line of the PDE menu button and then the PDE Specification line (Fig. 6.10) or click within the drawn shape (rectangle, in our case).

After the PDE Specification panel appears, we may mark the PDE type option required for solving the equation. In our case, it is the Elliptic option (default). The coefficients in this panel should be set by requiring the identity of our equation and the equation written in the panel. The equation appearing in the panel is written in a general vector form via the divergence and gradient $-div(c*grad(u))+au=f$; for scalar case it can be rewritten as $-\nabla(c\nabla(u))+au=f$. Therefore, the c, a, and f should be set as $c=1$, $a=0$, and $f=0$—Fig. 6.11.

Note: the set of parameters of the PDE Specification panel varies for different PDE types. These parameters may include initial conditions.

Step 5. At this stage, a triangular mesh should be generated to obtain a solution in their nodes. To achieve this. The Mesh Mode line in the popup menu of the Mesh button

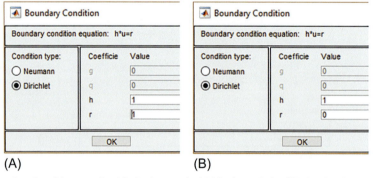

FIG. 6.9 The Boundary Condition panels with the inputted Dirichlet boundaries filled in for the example problem—horizontal (A) and vertical boundaries (B).

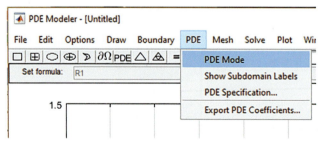

FIG. 6.10 The PDE mode selection.

should be selected. To see the node numbering, the Show Node Labels line should be also marked—Fig. 6.12.

The mesh can be changed by selecting the Refine Mesh or Jiggle Mesh line and returned to the starting view by selecting the Initialize Mesh line.

Step 6. Now the Solve PDE line of the popup menu of the PDE button should be selected—Fig. 6.13.

FIG. 6.11 The PDE Specification panels filled in for the PDE example.

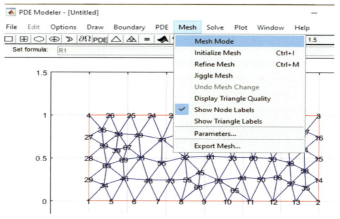

FIG. 6.12 The Mesh Mode selection; the Show Node Labels line is marked; triangle nodes is numbered as for the studied example.

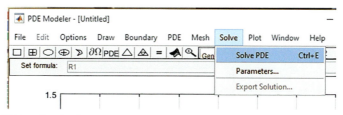

FIG. 6.13 The Solve PDE mode selection.

228 A MATLAB® Primer for Technical Programming in Materials Science and Engineering

When nondefault solver parameters, such as time or initial conditions, are required, the Parameters option can be selected. In such case, the Solve Parameters panel is opened. The default view of this window for a parabolic PDE is presented in Fig. 6.14.

In the case of nondefault initial conditions, they must be typed in the u(t0) field as a string line, i.e., 'sqrt(x.^2+y.^2)'.

For the discussed example, it is not necessary to introduce any changes in the Solve Parameters panel. Therefore, after selecting the Solve PDE mode, the resulting solution appears automatically. For our example, it looks as per Fig. 6.15.

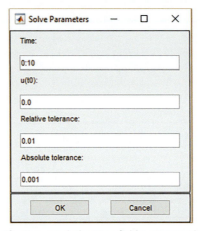

FIG. 6.14 The Solve Parameters specification panel; the Time field contents values 1, 2, ..., 10; u(t0) designate the initial condition fields.

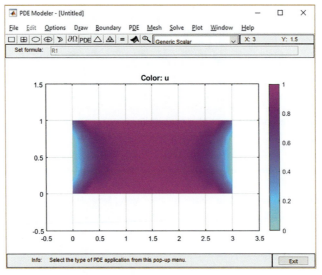

FIG. 6.15 View of the PDE solution in the PDE Modeler.

Step 7. To plot a 3D graph and/or provide some other graphical changes, the Parameters line in the popup menu of the Plot button should be selected—Fig. 6.16.

The Plot Selection window is opened—Fig. 6.17. To generate a 3D graph, the Height (3D) box should be marked. To add the five isotherm-lines to x,y plane of the graph, respectively, the Contour box can be marked. Type the five digit in the "Contour plot levels" field. The calculating mesh can be shown by marking the "Show mesh plot" box. If it is desirable to change the look-up colors, the needed map colors can be selected with the Colormap popup menu.

Finally, after selecting all the required parameters of the graph (as in the Fig. 6.17), click the Plot button of the Plot Selection window. The resulting plot is shown in Fig. 6.18.

Step 8 (optional). The PDE Modeler generates a program with commands, which carry out all of the steps described above. To save this program, select the "Save As" line of the popup File options of the main menu and type the desired file name in the File name field of the "Save As" panel that appears.

The PDE Modeler generates the file with the commands that carry out all of the solution steps. To save the program, select the Save As line of the popup menu of the main menu File option and type the name `ApExample6_5` in the File name field of the panel that appears.

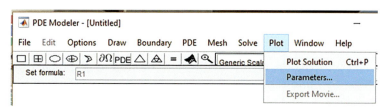

FIG. 6.16 Plot mode, the Parameters line selection.

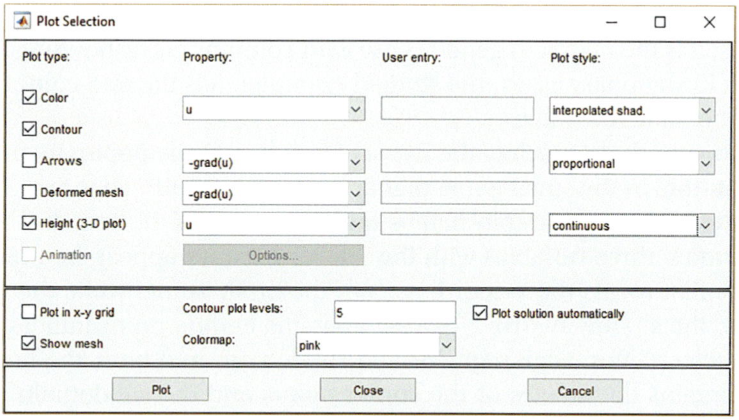

FIG. 6.17 The Plot Selection window; filled in for the example problem.

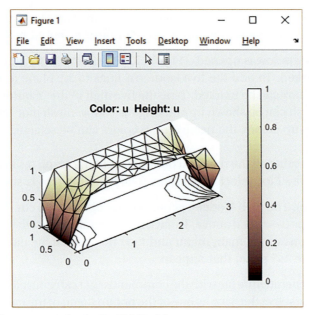

FIG. 6.18 The 3D solution representation by the PDE Modeler.

6.3.2 Exporting the obtained solution and mesh to the MATLAB® workspace

The PDE Modeler presents the results of the PDE solution in graphical form with a color bar. Nevertheless, the numerical values are frequently required. This can be achieved by exporting the defined u-values and coordinates of the mesh nodes at which these values were obtained to the MATLAB® workspace. To export the solution, select the Export Solution line in the popup menu of the Solution main menu button. In the small panel, that appears (Fig. 6.19A), choose to change/not change the default solution variable name and click the OK button. In the Workspace MATLAB® window, the variable with the selecting name appears; u is the matrix in general case with columns corresponding to the current time each but in stationary cases (the studied example), u is the one-column vector containing the 69 obtained u-values.

To export the mesh data, select the Export Mesh line in the popup menu of the Mesh main menu button. In the small panel that appears (Fig. 6.19B), choose to change or not change the default mesh variable names and click the OK button. In the Workspace MATLAB® window, three variables with the selecting names appear. In a two-row matrix p (points). the first row is the x-coordinates of the mesh points while the second row— y-coordinates; the six-row matrix e (edges) contains ending point indices, starting and ending parameter values, edge segment and subdomain numbers; the four-row matrix t (triangle) contains the indices of the corner points and the subdomain numbers. For practical needs, only the p data is important, but some commands may also use the e and t parameters. Thus, it is recommended to transfer all parameters to the workspace.

Chapter 6 • ODE-, PDEPE-solvers, and PDE Modeler tool   231

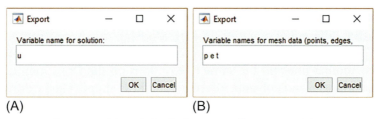

(A)   (B)

FIG. 6.19 Panels to export the PDE Modeler solution in the MATLAB® workspace; (A) Export obtained *u*-values, (B) export *x,y*-node coordinates and some other mesh parameters.

After exporting solution and mesh node coordinates, their values can be displayed; for example, with the following commands:

```
>> disp('    x    y    u'), disp([p' u])
        x           y          u
        0           0       0.5000
   3.0000           0       0.5000
   3.0000      1.0000       0.5000
...   (shortened)
```

6.3.3 Conversion of the solution from the triangular to rectangular grid

Exported values are obtained for the triangular mesh but, practically, the rectangular grid is of the greatest engineering interest. To transform the data from the triangle mesh nodes to orthogonal grid nodes, the `tri2grid` command can be used. The simplest form of the command is:

`U=tri2grid(p,t,u,x,y)`

where p, t, and u are, respectively, the coordinates, triangle parameters, and solution values, which should have been previously exported to the workspace (see the previous subsection); x and y are the coordinates of the rectangular grid nodes at which the *u* vales should be determined; and U is the determined *u*-values computed with the linear interpolation.

For our example:

```
>> x=0:0.5:3; y=0:0.25:1; U=tri2grid(p,t,u,x,y)
U =
  0.5000   1.0000 1.0000 1.0000 1.0000 1.0000 0.5000
       0   0.7987 0.9548 0.9826 0.9587 0.8018        0
       0   0.7302 0.9402 0.9746 0.9352 0.7277        0
       0   0.7992 0.9556 0.9828 0.9572 0.7895        0
  0.5000   1.0000 1.0000 1.0000 1.0000 1.0000 0.5000
```

Note, in the case of complicated plate geometry, some rectangular grid points can be located outside the plate, e.g., due to presence of a hole, rounding or cavity; these points are displayed as NaN (not a number) in the resulting *u*-values rectangular table.

6.3.4 Drawing in PDE Modeler

To solve a partial differential equation in PDE Modeler, it is necessary to first draw the actual object geometry. For this purpose, there are some basic shapes and appropriate buttons on the toolbar line and the same line options on the Draw menu. These shapes are an ellipse, rectangle, and polygon. Alternatively, these shapes can be drawn by entering commands in the Command Window. The button icons, their assignments, alternative commands, and examples are presented in Table 6.2.

Complicated geometry can be drawn with a combination of these shapes and correction of the expression in the Set formula field of the toolbar line. Some examples for various complicated objects are presented in Table 6.3.

Note, to quickly draw a square or circle using the Rectangle (centered) or Ellipse (centered) buttons, respectively, move the mouse pointer to the center point and drag the mouse keeping the right button down to any desired square point or the desired circle radius point.

6.4 Application examples

The commands in the examples below are written as user-defined functions with parameters and with short help part including a short definition of the function and the command for running it in the Command Window. Explanatory comments are not included into functions since it is assumed that the reader has gained enough experience after studying the previous chapters. All necessary explanations are given in text.

6.4.1 Examples with the ODE applications

6.4.1.1 Heat transfer with a temperature-dependent material property

A one-dimensional steady-state heat equation with heat source q and temperature-dependent thermal conductivity coefficient is described by the equation

$$k(T)\frac{dT}{dx} = q$$

where T is the temperature, K; x is the coordinate, m; q—heat flux (source), W/m^2; k—thermal conductivity of a material, W/(m K), which we assume to be dependent on temperature as follows:

$$k = 0.0004T - 0.05$$

Problem: Write a user-defined function with name `ApExample6_1` that calculates temperatures at x equal to 0, 0.02, ..., 0.12 m, displays the resulting table and generates graph $T(x)$, take starting, x_s, finishing, x_f and stepping, dx, coordinate values together with initial temperature $T_0 = 273$ K as input parameters of the function and the resulting two columns matrix with of the x and T values as the single output parameter. The function should contain the subfunction with ODE and $k(T)$ equations. Take the flux as $q = 12.7$ W/m^2.

Table 6.2 Draw buttons, commands, and examples of usage.

Button icon and name	Description	Alternative command	Example
Rectangle button	Draws a rectangle or square starting at a corner and dragging the mouse pointer (shown by +) to the diagonal corner.	`pderect([xmin xmax ymin ymax])` where `xmin`, `xmax`, `ymin`, and `ymax` are the coordinates of the diagonal corners.	Rectangle drawn with the button: Set R1 (figure showing rectangle R1). Alternatively: `>> pderect([-1 1 -0.5 0.5])`
Rectangle (centered) button	Draws a rectangle or square starting at the center.	The same as in previous case.	Square drawn with the button: Set SQ1 (figure showing square SQ1). Alternatively: `>> pderect([-0.5 0.5 -0.5 0.5])`

Continued

Table 6.2 Draw buttons, commands, and examples of usage—cont'd

Button icon and name	Description	Alternative command	Example
⬭ Ellipse button	Draws an ellipse or circle starting at the perimeter. The + (mouse pointer) are shown in the example at the end point.	pdeellip(xc, yc,a,b, phi) where xc, yc are the ellipse center coordinates, a and b are the ellipse semiaxes, phi— rotation angle in radian.	Ellipse drawn with the button: Alternatively: >> pdeellip(0,0,1,0.5,0)
Ellipse (centered) button	Draws an ellipse or circle starting at the center.	The same as in previous case or: pdecirc(xc, ys,r) where xc, yc, and r are the circle center coordinates and radius, respectively.	Circle drawn with the button: 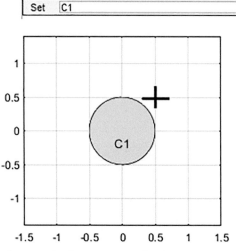 Alternatively: >> pdecirc(0,0,0.5)

Table 6.2 Draw buttons, commands, and examples of usage—cont'd

Button icon and name	Description	Alternative command	Example
Polygon button	Draws a polygon. Drag the mouse (shown by +) beginning from the starting point to the next point and so on; the last point should be the starting point.	pdepoly(X,Y) where X and Y are the vectors with coordinates of the each of the polygon corners (vertices).	Diamond drawn with the button: Alternatively: > > pdepoly([0 0.5 − 0.5],... [0 −0.5 −1 −0.5])

To solve the heat equation, substitute k and rewrite it in standard form:

$$\frac{dT}{dx} = \frac{q}{0.0004T - 0.05}$$

Now, the appropriate ODE command should be chosen. Unless otherwise indicated, it should be the `ode45` command.

The function `ApExample6_1` for the solution is:

Table 6.3 Complicated geometric shapes drawing in PDE Modeler.

What to draw	How to draw	Resulting figure
A shape with two subdomains: rectangle 2×0.4 and circle with $r=0.4$ with center at the (0.5,0) point.	• Mark the Grid and Snap options; use the default parameters for the axes limits and grid spacing. • Click the Rectangle (centered) button. Place the mouse pointer on point (−0,0) and drag to (1,1). • Click the Circle (centered) button. Place the mouse pointer on point (−0,0) and drag to (0,1) with the right mouse button pressed. • Check the expression in the Set Formula field—it should be R1+C1.	A view of the subdomains in the Draw mode
A 2×1 rectangular plate centered at (0,0) and with a radius of 0.2, rounded corners.	• Mark the Grid and Snap options and adjust the Axis limits, and Grid spacing as follows: x-axis to −1.5:0.1:1.5 and y-axis to −1:0.1:1. • Click the Rectangle button. Place the mouse pointer on point (−1,−0.5) and drag to (1,0.5). • Draw four circles with the radius 0.2 and the centers at (−0.8,−0.3), (−0.8,0.3), (0.8,−0.3), and (0.8,0.3)—with the ellipse (circled) button each. • Add four squares with the side 0.2, one in	

Table 6.3 Complicated geometric shapes drawing in PDE Modeler—cont'd

What to draw	How to draw	Resulting figure
	each corner. Enter the following expression R1-(SQ1+SQ2+SQ3+SQ4)+C1+C2+C3+C4 in the Set Formula field. • Select Boundary mode line in the Boundary menu option. • Select the Remove All Subdomain Borders line.	The two first plots show the plate view before and after activating the Boundary Mode. The lower plot represents the plate with rounded corners after marking the Remove All Subdomain Borders
A centered squared 2 × 2 plate with centered rectangular 0.2 × 1.2 cavity	• Mark the Grid and Snap options and adjust the Axis limits, and Grid spacing as follows: each of the x- and y-axis to −1.5 … 1.5 with spacing 0.1 • Click the Rectangle (centered) button. Place the mouse pointer to point (−0,0) and drag to (1,0.5). • Click the Rectangle (centered) button again. Place the mouse pointer on point (−0,0) and drag to (0.6,0.1). • Enter the following expression in the Set Formula field: SQ1-R1 • Select the Boundary mode line in the Boundary menu.	Set: SQ1-R1 The plot represents the plate with a cavity after activating the Boundary Mode.
L-shaped plate with a rounded corner. The sizes as per the second column.	• Mark the Grid and Snap options and adjust the Axis limits and Grid spacing as follows: x-axis to −1.5:0.1:1.5 and y-axis to −1:0.1:1.	

Continued

Table 6.3 Complicated geometric shapes drawing in PDE Modeler—cont'd

What to draw	How to draw	Resulting figure
	• Click on the polygon button and draw the polygon with the following (x,y) corner coordinates: (−1,−1), (−1,1), (0,1), (0,0), (1,0), (1,−1), and (−1,−1) to close this shape. • With the ellipse (circled) button, draw a circle with the radius 0.4 centered at point (0.4,0.4) each. With the rectangle (centered) button, draw the square at the center (0.2,0.2) and over to the Boundary mode. • Enter the following expression in the Set Formula field: P1+R1−C1. • Select the Remove All Subdomain Borders line.	

The two first plots in the right column show the plate view before and after activating the Boundary Mode. The lower plot represents the L-shaped plate with rounded corners after marking the Remove All Subdomain Borders.

```
function x_T=ApExample6_1(xs,xf,dx,y0)
%heat transfer with temperature dependent properties of material
%x_T=ApExample6_1(0,0.2,0.02,273)
[X,T]=ode45(@myODE,[xs:dx:xf],y0);
x_T=[X T];
plot(X,T(:,1))
xlabel('Coordinate'),ylabel('Temperature')
grid on
title('Heat transfer: temperature-dependent material property')
function dT=myODE(~,T)
A=-0.05;B=0.0004;
q=12.7;
k=A+B*T;
dT=q./k;
```

In this function, the resulted x and T column vectors are joined in the two-column x_T matrix by the x_T=[X T] command for further display. To run the function, the following command must be entered in the Command Window:

```
>>x_T=ApExample6_1(0,0.2,0.02,273)
x_T =
         0    273.0000
    0.0200    277.2301
    0.0400    281.3458
    0.0600    285.3559
    0.0800    289.2681
    0.1000    293.0893
    0.1200    296.8255
```

The resulting graph:

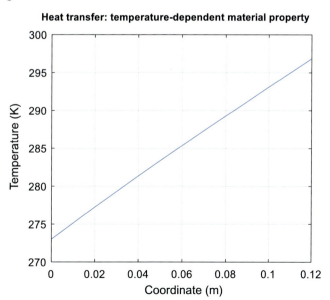

6.4.1.2 Polymeric materials kinetics; step-growth polymerization

The reaction rate of a catalyzed polyesterification is the first order differential equation:

$$-\frac{dc}{dt} = kc^2$$

where c is amount of substrate, t—time and k—polymerization rate constant.

Problem: Write a user-defined function with name `ApExample6_2` that calculates amount c of the COOH– component at the polymerizing times 0, 0.5, 1, 1.5, 2, 3, 5, 7, 9 and 11 s, displays the resulting table, and generates graph $c(t)$. Take t-span values together with the initial concentration $c_0 = 2$ mol as the input parameters of the function and the resulting two columns matrix as single output parameter. The matrix should contain the x and c values. The rate constant is $k = 2\,\text{s}^{-1}$.

The reaction rate equation in the ODE solver standard form is

$$\frac{dc}{dt} = -kc^2$$

For this equation solution, the `ode45` command is suitable as there is no contrary reason. It is advisable to use the simple extended form (see Section 6.1.2.2) when the ODE can be written strictly into the `ode45` command.

The function file for the problem solution is as below.

```
function t_c=ApExample6_2(tspan,y0)
%kinetics of a catalyzed polyesterification
%t_c=ApExample6_2([0:.5:2 3:2:11],2)
k=0.9;
[t,c]=ode45(@(t,c)-k*c.^2,tspan,y0);
t_c=[t c];
plot(t,c)
xlabel('Time, s'),ylabel('Amount of substrate, mol')
grid on
title('Polymeric materials. Catalyzed polyesterification')
axis tight
```

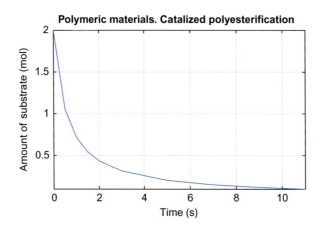

6.4.2 Examples with one-dimensional PDEs

6.4.2.1 One-dimensional diffusion equation with Neumann boundaries

Consider a simple diffusion equation describing the temporal and spatial changes of substance concentration u:

$$\frac{\partial u}{\partial t} = D\frac{\partial^2 u}{\partial x^2}$$

This equation, all its variables and initial conditions are the same as in Section 6.3.1 but, instead of the Dirichlet boundaries, the Neumann boundary conditions are assumed. This frequently is a more realistic case when the final boundary concentrations are not known in advance and/or, for example, the concentrations are uniform near the boundaries.

For this problem, the initial conditions:

$$u(x, 0) = \begin{cases} 1, & 0.45 \leq x \leq 0.55 \\ 0, & \text{elsewhere} \end{cases}$$

and the boundary conditions:

$$D\frac{\partial u(0, t)}{\partial x} = 0, \quad D\frac{\partial u(1, t)}{\partial x} = 0$$

Problem: Write a user-defined function with name `ApExample6_3` that calculates and generates a graph of the concentrations u along coordinate x from 0 to 1 and at t from 0 to 0.8. The input parameters of this function should be the m-number ($m=0$ for the Cartesian coordinates) and point numbers of the x and t values, and the output parameter—the two-dimensional matrix of each row of which is u-values along coordinate x at certain t. Take the x and t point numbers each equal to 20. To simplify the output, set the outputting u-matrix so that it has only five lines (No 1, 5, 10, 15, and 20) and six columns (No 1, 5, 10, 11, 15, and 20).

To solve the problem, the diffusion equation and initial and boundary conditions must be represented in standard form.

Rewriting the diffusion equation:

$$\frac{\partial u}{\partial t} = \frac{\partial}{\partial x}\left(D\frac{\partial u}{\partial x}\right)$$

and comparing with the standard form equations (presented early in Section 6.2), we obtain that they are identical when:

$$m = 0,$$

$$c\left(x, t, u, \frac{\partial u}{\partial t}\right) = 1,$$

$$f\left(x, t, u, \frac{\partial u}{\partial x}\right) = D\frac{\partial u}{\partial x},$$

$$s\left(x, t, u, \frac{\partial u}{\partial x}\right) = 0$$

The standard forms of the initial and boundary conditions are:

$$u_0 = \begin{cases} 1, & 0.45 \leq x \leq 0.55 \\ 0, & \text{elsewhere} \end{cases}$$

$0 + 1 \cdot D \dfrac{\partial u}{\partial x} = 0$ at $x = x_a = 0$, and $0 + 1 \cdot D \dfrac{\partial u}{\partial x} = 0$ at $x = x_b = 1$

To present the boundary conditions equations in the form required by the `pdepe` command, the following notations must be used:

$$p(0, t, u) = 0, \quad p(1, t, u) = 0$$

$$q(0, t) = q_a = 1, \quad q(1, t) = q_b = 1,$$

$$f\left(x, t, u, \frac{\partial u}{\partial x}\right) = D \frac{\partial u}{\partial x}$$

The commands for solving this problem are:

```
function u_table=ApExample6_3(m,n_x,n_t)
% Diffusion with Neumann boundaries
% To run: >> u_table=ApExample6_3(0,20,20)
xmesh=linspace(0,1, n_x);tspan= linspace(0,0.8, n_t);
u=pdepe(m,@mypde,@i_c,@b_c,xmesh,tspan);
u_table= u([1 5:5:n_t],[1 5 10 11 15 n_x]);
[X,T]=meshgrid(xmesh,tspan);
surf(X,T,u)
xlabel('Coordinate'),ylabel('Time'), zlabel('Concentration')
title('1D diffusion,Neumann conditions')
axis square
function [c,f,s]=mypde(x,t,u,DuDx)
% PDE
c=1;
D=0.1;
f= D*DuDx;
s=0;
function u0=i_c (x)
% initial condition
if x>=0.45&x<=0.55
  u0=1;
else
  u0=0;
end
function [pa,qa,pb,qb]=b_c(xa,ua,xb,ub,t)
% boundary conditions
pa=0;pb=0;
qa=1;qb=1;
```

After saving this program in the `ApExample6_3` file and entering the running command in the Command Window, the following table and plot appear.

```
>> u=ApExample6_3(0, 20, 20)
u_table =
        0        0  1.0000  1.0000       0        0
   0.0124  0.0659  0.2264  0.2264  0.0988  0.0124
   0.0587  0.0932  0.1524  0.1524  0.1086  0.0587
   0.0846  0.1002  0.1257  0.1257  0.1069  0.0846
   0.0962  0.1030  0.1142  0.1142  0.1060  0.0962
```

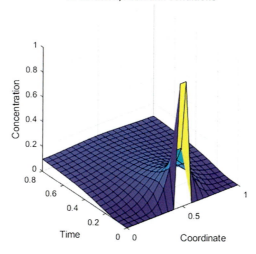

6.4.2.2 The diffusion Brusselator PDEs

The diffusion in a chemical reaction is described by the Brusselator-type partial differential equations:

$$\frac{\partial u}{\partial t} = 1 + u^2 v - 4u + \alpha \frac{\partial^2 u}{\partial x^2}$$

$$\frac{\partial v}{\partial t} = 3u - u^2 v + \alpha \frac{\partial^2 v}{\partial x^2}$$

where u and v are the substances concentrations and α—diffusion coefficient.

Problem: Write a user-defined function with name `ApExample6_4` that calculates u and v and generates two surface graphs in the same Figure window: $u(t,x)$ and $v(t,x)$ with coordinates x from 0 to 1 and at t from 0 to 1. Plot the graphs with the azimuth and elevation angles as −40 and 30 degrees, respectively. Take $\alpha = 0.02$ and the x and t point numbers each equal to 20. The `ApExample6_4` function should have two output parameters that display the matrix of the u and v concentrations each. To shorten the output, display only five lines and five columns (serial numbers 1, 5, 10, 15, and 20 for lines and columns) of the u and v matrices.

Solve the set of the PDEs with the following initial conditions:

$$u_0 = 1 + \sin 2\pi x, \quad v_0 = 3$$

and boundary conditions

$$\frac{\partial u(0, t)}{\partial x} = \frac{\partial u(1, t)}{\partial x} = 0 \text{ and } \frac{\partial v(0, t)}{\partial x} = \frac{\partial v(1, t)}{\partial x} = 0$$

To solve this set of equations, it is necessary to rewrite each of them in the standard form:

$$\frac{\partial u}{\partial t} = \frac{\partial u}{\partial x}\left(\alpha \frac{\partial u}{\partial x}\right) + 1 + u^2 v - 4u$$

$$\frac{\partial v}{\partial t} = \frac{\partial v}{\partial x}\left(\alpha \frac{\partial v}{\partial x}\right) + 3u - u^2 v$$

Alternatively, in the matrix form:

$$\begin{bmatrix} 1 \\ 1 \end{bmatrix} \cdot * \frac{\partial}{\partial t}\begin{bmatrix} u_1 \\ u_2 \end{bmatrix} = \frac{\partial}{\partial x}\begin{bmatrix} \alpha \frac{\partial u_1}{\partial x} \\ \alpha \frac{\partial u_2}{\partial x} \end{bmatrix} + \begin{bmatrix} 1 + u_1^2 u_2 - 4u_1 \\ 3u_1 - u_1^2 u_2 \end{bmatrix}$$

where u_1 and u_2 denote the u and v, respectively.

The initial and boundaries conditions are in the standard forms and rewritten as matrices as:

$$\begin{bmatrix} u_1(x, 0) \\ u_2(x, 0) \end{bmatrix} = \begin{bmatrix} 1 + \sin 2\pi x \\ 3 \end{bmatrix}$$

$$\begin{bmatrix} 0 \\ 0 \end{bmatrix} \cdot * \begin{bmatrix} u_1 \\ u_2 \end{bmatrix} + \begin{bmatrix} 1 \\ 1 \end{bmatrix} \cdot * \begin{bmatrix} \alpha \frac{\partial u_1}{\partial x} \\ \alpha \frac{\partial u_2}{\partial x} \end{bmatrix} = \begin{bmatrix} 0 \\ 0 \end{bmatrix}$$

The latter equation applies to both the a and b boundaries. In the `pdepe` command notations:

$c = [1; 1]; f = [0.02; 0.02]*DuDx; s = [1+u(1).^2*u(2) - 4*u(1); 3*u(1) - u(1).^2.*u(2)];$
$u_0 = [1 + \sin(2*pi*x); 3]; pa = 0; pb = 0; qa = 1; gb = 1.$

The commands for solving this problem are:

```
function [u_table,v_table]=ApExample6_4
% Diffusion- Brusselator ODEs
% To run: >>[u_table,v_table]=ApExample6_4
x=linspace(0,1,20);t=linspace(0,1,20);
m=0;
sol=pdepe(m,@myPDEs,@i_c,@b_c,x,t);
u=sol(:,:,1); v=sol(:,:,2);
u_table=u([1 5:5:20],[1 5:5:20],);v_table=v([1 5:5:20],[1 5:5:20],);
subplot(1,2,1)
surf(x,t,u)
view(-40,30)
title('u(t,x)'),xlabel('Coordinate'),ylabel('Time'),zlabel('u')
axis square tight
subplot(1,2,2)
surf(x,t,v)
view(-40,30)
title('v(t,x)'),xlabel('Coordinate'),ylabel('Time'),zlabel('v')
axis square tight
function [c,f,s]=myPDEs(x,t,u,DuDx)
%PDEs
c=[1;1];
alfa=0.02;
f=[alfa;alfa].*DuDx;
s=[1+u(1).^2.*u(2)-4*u(1);3*u(1)-u(1).^2.*u(2)];
function u0=i_c(x)
% initial conditions
u0=[1+sin(2*pi*x);3];
function [pa,qa,pb,qb]=b_c(xa,ua,xb,ub,t)
% boundary conditions
pa=[0;0];pb=[0;0];
qa=[1;1];qb=[1;1];
```

After saving this function in the `ApExample6_4` and running it, the following resulting data and graph appear:

```
>> [u_table,v_table]=ApExample6_4
u_table =
  1.0000 1.9694 1.1646 0.0034 1.0000
  1.7549 2.7059 1.3239 0.2239 0.5638
  2.4847 2.8247 1.5821 0.3430 0.4251
  2.3633 2.4080 1.6188 0.4165 0.3855
  1.9494 1.9509 1.4841 0.4765 0.3796
v_table =
  3.0000 3.0000 3.0000 3.0000 3.0000
  2.5067 1.8620 2.7642 3.0955 3.1380
```

```
1.5220  1.2152  2.3444  3.2611  3.3114
1.2533  1.2067  2.1234  3.3856  3.4732
1.3590  1.3718  2.1141  3.4621  3.6271
```

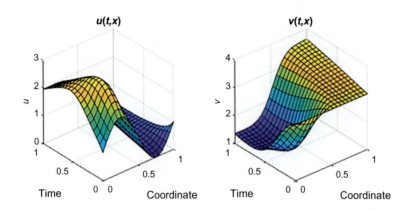

6.4.3 Examples with solving the two-dimensional PDEs with the PDE Modeler

6.4.3.1 Two-dimensional heat transfer equation with temperature-dependent property of a material

A transient 2D heat equation has the form:

$$\frac{\partial T}{\partial t} = \frac{\partial}{\partial x}\left(\alpha(T)\frac{\partial T}{\partial x}\right) + \frac{\partial}{\partial y}\left(\alpha(T)\frac{\partial T}{\partial y}\right)$$

where T is the temperature; x and y are the Cartesian coordinates; t—time; α is the ratio $k/(c_p\rho)$ and is called the thermal diffusivity of a material, k—thermal conductivity, c_p—heat capacity and ρ—material density. All variables in this equation are considered dimensionless.

We assume the thermal diffusivity is dependent on temperature as $\alpha = 0.3T + 0.4$.

Problem: Using the PDE Modeler, solve the heat equation for a rectangular 3 × 1 plate with Dirichlet boundaries: $T=1$ along two opposite horizontal sides of the rectangle and $T=0$ (default of the PDE Modeler) along the two opposite vertical sides. Set times $t=0, 1, 2, \ldots, 10$ and represent the final ($t=10$) temperature distribution $T(x,y)$ in a 3D plot. Transfer numbers in the workspace and display $T(x,y)$ table for the orthogonal grid at $x=0, 0.5, \ldots, 3$ and $y=0, 0.25, \ldots, 1$. Save the automatically created program in a file named ApExample6_5.

The heat equation is a parabolic type of the PDE. To match this equation with the required standard form (see Section 6.3), present the heat equation as

$$\frac{\partial T}{\partial t} - \nabla(\alpha(T)\nabla T) = 0$$

This equation is identical to the standard form when $u=T$, $d=1$, $c=\alpha(T)=0.3T+0.4$, and $a=f=0$.

First open the PDE Modeler window with the command

```
>> pdeModeler
```

In the popup menu of the Options menu button, mark the Grid and Snap lines and type the *x* and *y* limits as [−0.5 3.5] and [−0.5 1.5], respectively (after selecting the Axis Limits line). Use the general Application option—Generic Scalar.

Then, draw a rectangle with the Draw Mode by clicking the ▢ rectangle button. Now place the mouse arrow on point (0,0) and drag to point (3,1) of the plot. Check this rectangle geometry with the Object Dialog panel that appears after clicking within the rectangle.

Select the Boundary mode line of the popup menu of the Boundary main menu button and click on the appropriate boundary line to introduce 1 in the *h* and *r* fields into the Boundary Conditions panel for each of the horizontal sides of the rectangle. Check that the vertical rectangle lines have the required default conditions.

Now select the PDE Mode of the popup menu of the PDE main menu button, mark the Parabolic type of PDE and type: 0.4+0.04*u within the *c* field; 0 within the *a* and *f* fields and 1 within the *d* field.

Initialize the triangle mesh in the Mesh Mode (a line in the popup menu of the Mesh button).

Select the Parameters line in the popup menu of the Solve main menu button and check the Time field of the Solve Parameters panel. 0:10 (as it is the problem start and final times) should be written. After that, click the Solve PDE line of the popup menu of the Solve mode button. The 2D solution with colored bar appears as below.

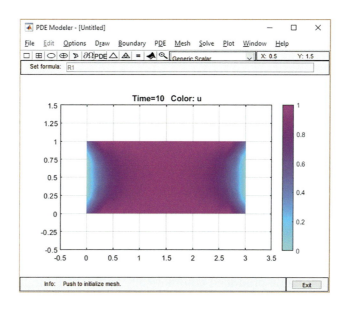

Mark "Height(3D plot)" and "Show mesh" options of the Parameters panel, which appears after selecting the appropriate line in the Plot menu button. The following Figure window appears with the 3D plot of the solution:

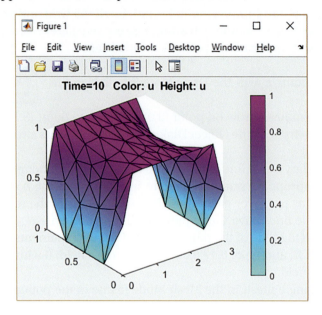

To transfer the solution and mesh parameters to the MATLAB® workspace, select the Export Mesh and Export Solution lines within the popup menus of the respective Solve and Mesh menu buttons. If the variable names are not changed, the solution in the u matrix (each matrix page corresponds to the given times, i.e., $u(:,:,1)$ corresponds to $t=1$ and $u(:,:,11)$—to $t=0.08$) while mesh parameters are within the p, e, and t matrices. After this, to obtain u-values at the required orthogonal grid points, the following commands should be entered with the resulting table displayed in the Command Window.

```
>> x=0:.5:3;y=0:.2:1;
>>T=tri2grid(p,t,u(:,end),x,y)
T =
    0.5000  1.0000  1.0000  1.0000  1.0000  1.0000  0.5000
         0  0.8037  0.9561  0.9831  0.9599  0.8066       0
         0  0.7364  0.9419  0.9754  0.9370  0.7338       0
         0  0.8041  0.9569  0.9833  0.9585  0.7945       0
    0.5000  1.0000  1.0000  1.0000  1.0000  1.0000  0.5000
```

The PDE Modeler generates the file with commands that carry out all steps of the solution. To save the program, select the Save As line of the popup File menu and type the name ApExample6_5 in the File name field of the panel that appears. After this, open the saved file in the MATLAB® Editor and write `ApExample6_5` in the function definition line instead of the default name produced automatically by the PDE Modeler.

6.4.3.2 Two-dimensional transient diffusion equation with coordinate-dependent initial conditions

The transient 2D diffusion equation has the form:

$$\frac{\partial u}{\partial t} = D\left(\frac{\partial^2 u}{\partial x^2} + \frac{\partial^2 u}{\partial y^2}\right)$$

where u is the concentration of a substance; x and y are the Cartesian coordinates; t—time; D is the material-dependent diffusion coefficient, which can be assumed 1 for simplification. All variables in this equation are considered dimensionless.

Problem: Using the PDE Modeler, solve the diffusion equation for a squared plate of 1×1 with the Neumann boundaries, $du/dt = 0$ along all the square sides and with the following initial concentrations:

$$u_0 = \begin{cases} 1, & at\ 0.4 \leq x \leq 06, 0.4 \leq y \leq 0.6 \\ 0, & in othercase \end{cases}$$

Set times $t = 0, 0.01, \ldots, 0.08$ (rapid-flowing diffusion) and represent the final ($t = 0.08$) concentration distribution $u(x,y)$ in 3D plot, transfer numbers in the workspace and display the concentration table for the orthogonal grid with x and y equal to 0, 0.25, …, 1 each. Save the automatically created program in the file named ApExample6_6.

To solve the problem, open the PDE Modeler window with the command

```
>> pdeModeler
```

In the popup menu of the Options menu button, mark the Grid and Snap lines and type the x and y limits as [−0.5 1.5] each (after selecting the Axis Limits line). Set the Grid spacing with step 0.25 and mark the Axis Equal line. Select the general Application option—Generic Scalar.

To draw the square, click this ▭ rectangle button, place the mouse arrow on the point (0,0) and drag to point (1,1) of the plot. Check this square geometry with the Object Dialog panel, which appears after clicking within the rectangle.

Select the Boundary mode line of the popup menu of the Boundary main menu option and click on the appropriate boundary line. After the Boundary Conditions panel appears, mark the Neumann condition box and enter 0 in the g and q fields; do this for all four square sides.

Now select the PDE Mode of the popup menu of the PDE main menu option, select the PDE Specification and the Parabolic type of PDE found there and type 1 within the c and d fields and 0 within the a and f fields.

Initialize the triangle mesh in the Mesh Mode (a line in the popup menu of the Mesh button). Select the Parameters line in the popup menu of the Solve main menu option and type 0:0.1:0.08 in the Time field of the appeared Solve Parameters panel. In the same panel, the following string with initial conditions should be typed in the u(t0) field: '(x>0.4&x<0.6)&(y>0.4&y<0.6)'. The result of this logical operation is u0 matrix of logical 1s/0s with u0=1 when *x* and *y* are within the required ranges and u0=0 out of the range; the logical values are used as conventional numbers. After inputting the initial conditions, click the Solve PDE line of the Solve button and select the Jet color in the Plot Selection panel selected with the Plot main menu option. The 2D solution with colored bar appears as below:

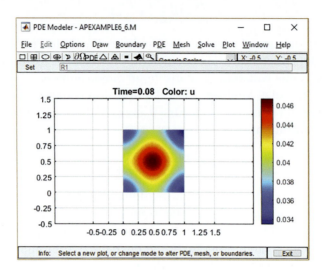

Marking the Height(3D) box the 3D plot is generated as below.

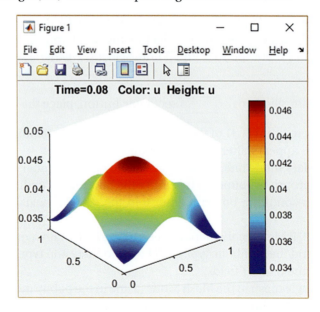

To transfer the solution and mesh parameters to the MATLAB® workspace, select the Export Mesh and Export Solution lines within the popup menus of the respective Mesh and Solve menu buttons. If the variable names were not changed, the solution is in the u matrix (each matrix page corresponds to the given times, i.e., u(:,:,1) corresponds to $t=0$ and u(:,:,11) to $t=0.08$) while the mesh parameters are within the p, e and t matrices. After this, to obtain the *u*-values at the required orthogonal grid points, the following commands must be entered, with the resulting table displayed in the Command Window.

```
>> x=0:.25:1;y=0:.25:1;uxy=tri2grid(p,t,u(:,end),x,y)
uxy =
    0.0351  0.0378  0.0403  0.0369  0.0338
    0.0381  0.0409  0.0435  0.0399  0.0367
    0.0410  0.0440  0.0467  0.0428  0.0394
    0.0381  0.0409  0.0433  0.0397  0.0364
    0.0351  0.0377  0.0400  0.0366  0.0335
```

Any asymmetry in the concentration values is the result of the triangulation of the square plate and subsequent interpolation from the triangle to the orthogonal grid.

To save the automatically generated program, select the Save As line of the popup File menu and type the name ApExample6_6 in the File name field of the panel that appears. After this, open the saved file in the MATLAB® Editor and write `ApExample6_6` in the function definition line instead of the default name produced automatically by the PDE Modeler.

6.5 Questions and exercises for self-testing

1. Which command of the ODE solver is recommended for an ordinary differential equation, when the class of the problem is unknown in advance? Choose the correct answer: (a) `ode23`, (b) `ode45`, (c) `ode15s`, (d) `ode113`.
2. For an ODE that, as expected, has solution with singularity the following command should be tried first: (a) `ode15s`, (b) `ode15i`, (c) `ode23s`, (d) `ode23tb`. Choose the correct answer.
3. The `tspan` vector specifying the interval of solution in the ODE solvers should have: (a) at least one value of the starting point of the interval; (b) at least two values—the starting and end points of the interval; (c) several values within the interval, including the starting and end points; (d) the two latter answers are correct. Choose the correct answer.
4. In the `pdepe` solver, a vector `xmesh` specifies: (a) *x* and *t* coordinates for solving partial differential equation; (b) *x*-coordinates of the points at which a solution is required for every time value; (c) the *t*-coordinates of the points at which a solution is required for every *x* value. Choose the correct answer.
5. How many and which subfunctions should be written for the `pdepe` command within the function file: (a) one function with PDE, (b) two functions—with PDE and with the initial condition, (c) three functions—with PDE, with the initial and boundary conditions. Choose the correct answer.

6. A PDE should be solved with Neumann boundary conditions $k(du/dx)=0$ at each of a and b boundaries. The pa, pb, qa, and qb arguments in the boundary conditions function should be written so: (a) pa=ua; pb=ub; qa=0; qb=0; (b) pa=0; pb=0; qa=0; qb=0; (c) pa=0; pb=0; qa=1; qb=1; (d) pa=0; pb=0; qa=k; qb=k. Choose the correct answer.

7. The following dimensionless PDE should be solved with the pdepe command: $c_p \rho \frac{\partial T}{\partial t} = \frac{\partial}{\partial x}\left(k \frac{\partial T}{\partial x}\right)$, where $c_p = 0.9$, $\rho = 1.01$, and $k = 0.95$ are material-dependent constants, T—temperature and t—time. Within the subfunction of the function file, this equation should be written as: (a) c=0.9*1.01; f=0.95*DuDx; s=0; (b) c=0.9*1.01; f=D(0.95*DuDx); s=0; (c) c=0.9; f=0.95*DuDx; s=1.01; (d) c=0.9/0.95; f=DuDx; s=1.01/0.95. Choose the correct answer.

8. One of the boundaries given as $du/dt = 5$ for a two dimensional PDE. What should be marked and typed into the Boundary Condition panel of the PDE Modeler's Boundary mode? (a) Mark the Dirichlet boundary box and type $h=1$, $r=5$ in the appropriate fields; (b) Mark the Neumann boundary box and type $q=0$, $g=5$ in the appropriate fields; (c) mark the Dirichlet boundary box and type $h=5$, $r=1$ in the appropriate fields; (d) Mark the Neumann boundary box and type $q=5$, $g=0$ in the appropriate fields. Choose the correct answer.

9. The following dimensionless PDE should be solved with the PDE Modeler:
$\frac{\partial T}{\partial t} = \frac{\partial}{\partial x}\left(D \frac{\partial T}{\partial x}\right) + \frac{\partial}{\partial y}\left(D \frac{\partial T}{\partial y}\right)$, where $D = 0.7$ is the material-dependent constant, T—temperature, t—time. What must be marked and typed into the PDE specification panel of the PDE Modeler's PDE mode?

(a) Mark the "Elliptic" type box and enter 0.8, 0, and 0 in the c, a, and f fields, respectively;
(b) Mark the Parabolic type box and enter 0.8, 0, 0, and 1 in the c, a, f, and d fields, respectively;
(c) Mark the Hyperbolic type box and enter 0.8, 0, 0, and 1 in the c, a, f, and d fields, respectively;
(d) Mark the Eigenmodes type box and enter 0.8, 0, and 1 in the c, a, and d fields, respectively.

Choose the correct answer.

10. A two-dimensional parabolic type PDE that should be solved with the PDE Modeler for the 2 × 1 rectangular plate, has the following initial conditions: $u_0 = 1$ at $x >= 1$ and $u0 = 0$ in other case. In which mode, where, and what should be inputted to realize these initial conditions?

(a) PDE mode, the PDE specification panel, mark the Parabolic type equation and type if x<=1,u0=0;else u0=1; end;
(b) Solve PDE mode when the parabolic type equation was selected previously, the Solve parameters panel, type the "x>=1" string in the u(t0) field;
(c) Solve PDE mode (when the parabolic type equation where selected in an one of the previously mode) the Solve parameters panel, type the "if x<=1, u0=0; else u0=1;" string in the u(t0) field;

(d) both (b) and (c) answers are correct.

Choose the correct answer.

11. The time-change of a radioactive material amount u describes by the following ODE: $\frac{du}{dt} = -ku$, where t—time and k is a constant representing radioactive properties of the material. Solve this equation with the ODE solver with the times $t = 0, 2, 4, \ldots, 20$ day and with the $k = 0.135 \text{ day}^{-1}$ and the initial material amount $u0 = 500$ g. Write for this a user-defined function with name `Problem_6_11` that has starting t_s, final t_f and step dt times and initial u_0 values as the input parameters and resulting matrix `t_u` with t and u columns as output parameter. Use the simplest possible form for presentation of the solving ODE in the ODE command. The function should represent the $u(t)$ plot with grid, axes labels and a caption.

12. The one-dimensional time-independent Schrödinger-type equation for single particle moving has the following dimensionless form: $-\frac{\partial^2 Y}{\partial X^2} + VY = EY$ where Y is stationary wave function, E is the total energy (an odd integer value constant), V—potential energy that for the "harmonic oscillator" case is $V = X^2$ and X is the coordinate. Write a user-defined function with the name `Problem_6_12` without output parameters that solves the equation with the `ode45` command and has starting x_s and final x_f times and initial y_0 values as the input parameters. Take $x_s = 0$, $x_f = 1.5$, $y(0) = 0$, $y'(0) = 1$. The results should be represented in graph of the $Y^2/(2\int_{x_s}^{x_f} Y^2 dx)$ as a function of X for the four curves obtained at different E constants: $E = 1, 3, 5,$ and 7. Add the grid, axes labels, a caption and legend to the plot. The $\int_{x_s}^{x_f} Y^2 dx$ should be defined with the `trapz` command.

13. The radiative cooling of a metallic body describes by the following ODE:

$$\frac{dT}{dt} + k(T^4 - T_s^4) = 0$$

where T—temperature, °C; T_s—surrounding temperature; k—object-dependent constant, C³/min, (represents generally the set material-dependent properties— $\varepsilon\sigma/(\rho c\tau)$ that are constants here). Solve this equation with the ODE solver at $t = 0, 0.5, 1, 2, 3, 5, 8, 11, 20, 25$ min and for the $k = 3.1 \cdot 10^{-8} \text{ min}^{-1}$, and $T_s = 20°C$ and initial temperature $T_0 = 500°C$. Write for this a user-defined function with the name `Problem_6_13`, which contains the time vector, surrounded temperature and initial u_0 values as the input parameters and the resulting matrix `t_T` with t and T columns as the output parameter. Use the simplest possible form for presentation of the solving ODE in the ODE command. The function should represent the $T(t)$ plot with grid, axes labels and a caption.

14. For some mathematical models of diffusion, a set of two reaction-diffusion equations is used:

$$\frac{\partial u}{\partial t} = D_u \frac{\partial^2 u}{\partial x^2} - \frac{1}{1+v^2}$$

$$\frac{\partial v}{\partial t} = D_v \frac{\partial^2 v}{\partial x^2} + \frac{1}{1+u^2}$$

where u and v are participating substance concentrations; $D_u=0.5$ and $D_v=0.45$ are the material-dependent diffusion coefficients of the u- and v-substances, respectively, x is the Cartesian coordinate that corresponds to $m=0$ in the `pdepe` command; t—time. The x varies from 0 to 1, and t—from 0 to 0.3; x and t point numbers are 20 each. All parameters are dimensionless. The initial conditions are assigned as $u(x,0)=1+(1/2)\cos(2\pi x)$ and $v(x,0)=3$. The boundary conditions are $\frac{\partial u(0,t)}{\partial x}=\frac{\partial u(1,t)}{\partial x}=0$ and $\frac{\partial v(0,t)}{\partial x}=\frac{\partial v(1,t)}{\partial x}=0$.

Write a user-defined function with name `Problem_6_14` that solves with the `pdepe` command the given set of the PDEs with its initial and boundary conditions. The function has no output parameters and contains the following input parameters: starting times, coordinates, final times and coordinates as well as point numbers for the time and coordinate meshes. The resulting concentrations as function of time and coordinate should be represented in two plots of the same Figure window: one plot with $u(t,x)$ and another with $v(t,x)$.

15. The one-dimensional heat transfer equation with temperature-dependent properties of a material has the form:

$$\frac{\partial T}{\partial t}=\frac{\partial}{\partial x}\left(\alpha(T)\frac{\partial T}{\partial x}\right)$$

where T—temperature, x and t are the coordinate and time, respectively, and α—is the thermal diffusivity coefficient that changes with temperature as $\alpha=0.3T+0.4$. The initial and boundary conditions are as follows: $T(x,0)=x$ (spatially linear temperature-change from 0 to 1), $T(0,t)=0.1$, $T(1,t)=0.2$. The x varies from 0 to 1, and t—from 0 to 0.3; all parameters are dimensionless.

Write a user-defined function with name `Problem_6_15` that solves with the `pdepe` command the heat equation with the given initial and boundary conditions. The function has no output parameters and has the following input parameters starting times and coordinates, final time and coordinates as well as point numbers for the time and coordinate meshes; take 20 points for each mesh. The problem is to be solved using a Cartesian coordinate with $m=0$ in the `pdepe` command. The resulting temperatures as function of time and coordinate must be represented in the $T(t,x)$ plot generated in the reverse y-view (use the `axis ij` command for this).

16. The diffusion a substance in case of the one-dimensional spherical coordinate system is described by the following equation:

$$\frac{\partial u}{\partial t}=D\frac{1}{r^2}\frac{\partial}{\partial r}\left(r^2\frac{\partial u}{\partial r}\right)$$

where u—a substance concentration, r—radial coordinate and D—diffusion coefficient, which is assumed to be constant. All parameters are dimensionless. The r is changed in the 0 ... 1 range, and t—in the 0–1.1 range.

The initial and boundary conditions are: $u(r,0)=u0$, $du(0,t)/dr=0$, $du(1,t)=0$. Write a user-defined function with name `Problem_6_16` that solves with the `pdepe` command the heat equation with the given initial and boundary conditions. The function has no output parameters and has the following input parameters; starting times and coordinates, final time and coordinate, and number mesh points for time and coordinate; take 20 points for each mesh. The problem is solving in spherical coordinate with $m=2$ in the `pdepe` command. Resulting concentrations as function of time and radial position should be represented in the $u(t,r)$ plot generated in the reverse y view (use the `axis ij` command for this).

17. The Poisson equation has many applications in the materials science (potential of ions, electron-electron interaction in metal-insulator semiconductors, electron density, etc.). Its two-dimensional dimensionless form is:

$$-\left(\frac{\partial^2 u}{\partial x^2} + \frac{\partial^2 u}{\partial y^2}\right) = f$$

where u is the application-dependent searching function, x and y—Cartesian coordinates, f—application-dependent constant. The equation should be solved with the PDE Modeler for the ring-shaped plate of outer and inner circles of radii 6 and 1, respectively. Select the Generic Scalar line in the Options application mode. Take $f=10$ and the Dirichlet boundaries are $u=0$ on the all ring sides and $f=10$. Choose the refined triangle mesh. The results should be represented in the 2D and 3D plots. Each of them should show the five contour levels and the "jet" color map as well as contain the color bar. The resulting u and triangle p, e, t-values should be exported to the MATLAB® workspace; display u at the orthogonal mesh with the `tri2grid` command when the x and y coordinates are equal to $-3, -2.5, \ldots, 3$ each.

18. The dimensionless heat equation with temperature dependent properties of material:

$$\frac{\partial T}{\partial t} = \frac{\partial}{\partial x}\left[\alpha(T)\frac{\partial T}{\partial x}\right] + \frac{\partial}{\partial y}\left[\alpha(T)\frac{\partial T}{\partial y}\right]$$

(see nomenclature in application example 6.4.3.1) should be solved with the PDE Modeler for a squared 6×6 plate (the center 0,0) with rectangular 1×1.2 cavity in the center of the plate. Select the Generic Scalar line in the Options application mode. The thermal diffusivity is $\alpha = 0.3\,T + 0.4$. The Neumann boundaries $\frac{\partial T}{\partial x} = \frac{\partial T}{\partial y} = 0$ are on the outer plate sides while the Diriclet boundaries are on the cavity borders. The initial conditions are $T=1$ at $-0.5 \leq x \leq 0.5$ and $-2.5 \leq y \leq -1.5$. Set times in range 0 … 0.1 with step 0.01. Choose the `refined` options for the triangle mesh. Represent the final u-values in the 2D and 3D plots, which should show the four contour levels and "jet" color map as well as contain the color bar. The resulting u and triangle p, e, t-values should be exported to the MATLAB® workspace; transform u at $t=0.1$ at the rectangular mesh with the `tri2grid` command when the x and y coordinates are equal to $-3, -2, \ldots, 3$ each and display five first lines with five columns each.

19. The transverse vibrations in the 2D plane describe by the wave equation

$$\frac{\partial^2 u}{\partial t^2} = \frac{d}{dx}\left(\frac{du}{dx}\right) + \frac{d}{dy}\left(\frac{du}{dy}\right)$$

where u is the wave perturbation, x, y, and t are the horizontal, vertical coordinates, and time, respectively. Consider this equation with the PDE Modeler for 2×2 square membrane (the center at 0,0), which is fixed at the two horizontal sides and is free at the two vertical sides. Select the Generic Scalar line in the Options application mode. For both fixed sides, the Dirichlet, $u=0$, and, for the free sides, the Neumann boundaries, $u'=0$ are specified. Initial conditions are $u_0=\mathrm{atan}(\cos(\mathrm{pi}/2*y))$ and $u'_0=3*\sin(\mathrm{pi}*y).*\exp(\sin(\mathrm{pi}/2*x))$. Set 31 equally spaced values of the times in range 0 … 5. Represent the final u-values in the 2D and 3D plots that should show the four contour levels and the "summer" color map as well as contain the color bar. Resulting u and triangle p, e, t-values should be exported to the MATLAB® workspace; display u at $t=0.1$ at the orthogonal mesh with the `tri2grid` command when the x and y coordinates are equal to -1, -2.5, …, 1 each.

6.6 Answers to selected questions and exercises

2.(a) `ode15s`.
4.(b) x-coordinates of the points at which a solution is required for every time value.
6. (c) `pa=0; pb=0; qa=1; qb=1`.
8.(b) Mark the Neumann boundary box and type $q=0$, $g=5$ in the appropriate fields.
10.(b) Mark the Parabolic type box and enter 0.8, 0, 0, and 1 in the c, a, f, and d fields, respectively.
12.

```
>> Problem_6_12([1:2:7],0,1.5,[0;1])
```

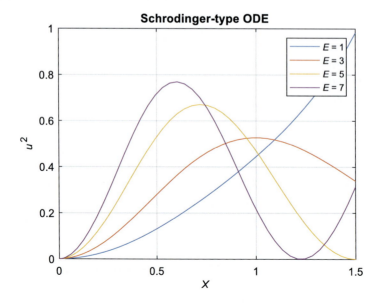

14.

```
>> Problem_6_14(0,1,0,0.3,20,20)
```

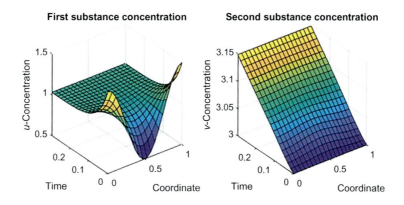

16.

```
Problem_6_16(0,0.4,0,1,20,20)
```

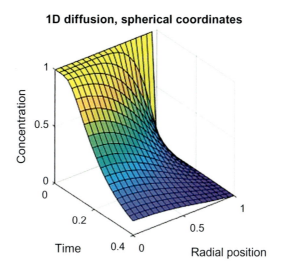

18.

```
>> Problem_6_18
```

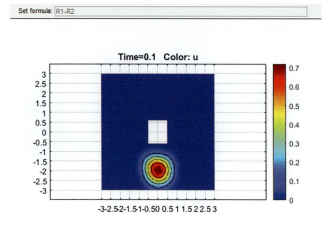

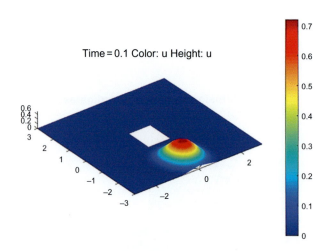

```
>> x=-3:3;y=x;u_table=tri2grid(p,t,u(:,end),x,y);T=u_table(1:
end-2,2:end-1)
    T =
    -0.0000  -0.0010   0.1252   0.0044   0.0000
     0.0000   0.0530   0.7168   0.0630   0.0000
     0.0000   0.0061   0.0800   0.0037   0.0000
     0.0000   0.0000      NaN   0.0000   0.0000
    -0.0000  -0.0000  -0.0000   0.0000  -0.0000
```

Appendix: Characters, operators, and commands for mastering programs

The special characters, variables, operators, and commands presented in this book are listed in the tables below.

Table A.1 Special characters and operators used for arithmetic, array, and matrix manipulations.

No	Operator or character	Description	Alternative command or character name	Page/s
1	+	Addition	plus	18, 48
2	−	Subtraction	minus	18, 48
3	*	Scalar and matrix multiplication	mtimes	18, 49
4	.*	Element-wise multiplication	times	55
5	/	Right matrix division	mrdivide	18, 53
6	./	Element-wise right division	rdivide	55
7	\	Left matrix division	mldivide	18, 53
8	.\	Element-wise left division	ldivide	55
9	^	Exponentiation, matrix power	mpower	18
10	.^	Element-wise power	power	55
11	=	Specifies a value/set of values to a variable	Assignment	17
12	%	Is used for comments and specifies an output number format	Percent	17, 33
13	( )	Is used for input arguments, sets precedence, addresses the vector/matrix/array elements	Parentheses	8, 45
14	[ ]	Is used for input of the vector/matrix/array elements	Brackets	39
15	(space)	Separates elements of vectors/arrays/matrices, and used also into output function specifications	Space	32, 39
16	:	Creates vectors, used also for loops, iterations, and for selecting all array elements	Colon	41, 45, 86
17	,	Separates commands on the same line and elements into arrays	Comma	18, 39
18	;	Suppresses displaying if wrote after the command; separates commands on the same line and marix rows	Semicolon	17, 45
19	...	Denotes that a long statement is to be continued on the next line	Ellipsis	17

Continued

Table A.1 Special characters and operators used for arithmetic, array, and matrix manipulations—cont'd

No	Operator or character	Description	Alternative command or character name	Page/s
20	'	Transposes vector, array, or matrix and is used for a text string generation	Apostrophe	32, 43, 46
21	~	Uses to ignore function argument or output; is used as logical not	Tilde	355

Table A.2 Relational and logical operators.

No	Operator	Description	Alternative command	Page
1	==	Equal (double); element-wise	eq	76
2	>	Greater than; element-wise	gt	76
3	>=	Greater than or equal to; element-wise	qe	76
4	<	Less than; element-wise	lt	76
5	&	Logical AND; element-wise	and	78
6	~	Logical NOT	not	78
7	|	Logical OR; element-wise	or	78
8	<=	Less than or equal to; element-wise	le	76
9	~=	Not equal; element-wise	ne	77

Table A.3 Predefined variables.

No	Variable	Description	Page
1	ans	Last calculated value	17
2	eps	Smallest difference between two numbers	28
3	i	$\sqrt{-1}$, the same as j	28
4	inf	Infinity	28
5	j	$\sqrt{-1}$, the same as i	28
6	NaN	Not a number	28
7	pi	Number π	21

Table A.4 Alphabetical list of nongraphical commands.

No	Command	Description	Page/s
1	abs	Absolute value	19
2	acos	Inverse cosine for angle in radians	19
3	acosd	Inverse cosine for angle in degrees	19
4	acot	Inverse cotangent for angle in radians	19
5	acotd	Inverse cotangent for angle in degrees	19
6	asin	Inverse sine for angle in radians	19
7	asind	Inverse sine for angle in degrees	19

Table A.4 Alphabetical list of nongraphical commands—cont'd

No	Command	Description	Page/s
8	atan	Inverse tangent for angle in radians	19
9	atand	Inverse tangent for angle in degrees	19
10	ceil	Rounds towards infinity	19
11	clc	Clear the command window	17
12	clear	Remove variables from the Workspace	29
13	convdensity	Converts density units	23
14	cos	Cosine for angle in radians	19
15	cosd	Cosine for angle in degrees	20
16	cosh	Hyperbolic cosine	20
17	cot	Cotangent for angle in radians	20
18	cotd	Cotangent for angle in degrees	20
19	coth	Hyperbolic tangent	20
20	cross	Calculates cross product of a 3D vector	60
21	det	Calculates a determinant	60
22	diag	Creates a diagonal matrix from a vector	60
23	diff	Calculates a difference, approximates a derivative	237
24	disp	Display output	32
25	doc	Displays HTML documentation in the Help window	24
26	dot	Calculates scalar product of two vectors	60
27	edit	Opens Editor window	204
28	else, elseif	Is used with if; conditionally executes if statement condition	84, 85
29	end	Terminates scope of for, while, if statements, or serves as last index	43, 84, 86
30	erf	Error function	20
31	exp	Exponential	20
32	eye	Creates a unit matrix	52
33	factorial	Factorial function	20
34	find	Finds indices of certain elements of array	80
35	fix	Rounds towards zero	20
36	floor	Rounds off toward minus infinity	20
37	fminbnd	Finds a function minimum within the interval	228
38	fminsearch	Multidimensional optimization, fitting	241
39	for	Is used to repeat execution of command/s	86
40	format	Sets current output format	29
41	fprintf	Display formatted output	34
42	function	Defines a new function	215
43	fzero	Solves a single-variable equation	225
44	gamma	Gamma function	20
45	global	Declares a global variable	218
46	help	Displays explanations in the Command Window	23
47	if	Conditionally execute	84
48	input	Prompts to user input	211
49	integral	Calculates integral	241
50	interp1	One-dimensional interpolation	224
51	inv	Calculates the inverse matrix	53
52	length	Number of elements in vector	61

Continued

Table A.4 Alphabetical list of nongraphical commands—cont'd

No	Command	Description	Page/s
53	linspace	Generates a linearly spaced vector	41
54	log	Natural logarithm	20
55	log10	Decimal logarithm	20
56	lookfor	Search for the word in all help entries	24
57	max	Returns maximal value	61
58	mean	Calculates mean value	61
59	median	Calculates median value	61
60	min	Returns minimal value	61
61	num2str	Converts numbers to a string.	61
62	ode113	Solves nonstiff ODEs	363
63	ode15s	Solves stiff ODEs	350
64	ode23	Solves nonstiff ODEs	363
65	ode23s	Solves stiff ODEs	363
66	ode23t	Solves stiff ODEs	364
67	ode23tb	Solves stiff ODEs	364
68	ode45	Solves nonstiff ODEs	350
69	odeset	Sets ODE options	359
70	ones	Creates an array with ones	58
71	pchip	Interpolates by the Piecewise Cubic method	240
72	pdepe	Solves 1D (spatially) PDE	368
73	polyder	Evaluates derivative of the polynomial	241
74	polyfit	Evaluates the polynomial value	286
75	polyval	Numerical integration with the	287
76	quad	Calculates integral, Simpson's rule	232
77	quadl	Calculates integral, Lobatto quadrature	241
78	quadgk	Calculates integral, Gauss-Kronrod quadrature	241
79	quad2d	Calculates double integral	241
80	rand	Generates an array with uniformly distributed random numbers	58
81	randi	Generates an array with integer random numbers from uniform discrete distribution	62
82	randn	Generates an array with normally distributed numbers	58
83	repmat	Duplicates a matrix	62
84	reshape	Changes size of array/matrix	62
85	rng	Controls random number generator	63
86	roots	Defines the roots of a polynomial	240
87	round	Rounds off toward nearest decimal or integer	21
88	sin	Sine	21
89	sind	Sine for angle in degrees	21
90	sinh	Hyperbolic sine	21
91	size	Size of vector/array/matrix	62
92	sort	Arranges elements in ascending or descending order	62
93	spline	Interpolates with the cubic spline	240
94	sqrt	Square root	21
95	std	Calculates standard deviation	62
96	strvcat	Concatenates strings vertically	63

Table A.4 Alphabetical list of nongraphical commands—cont'd

No	Command	Description	Page/s
97	sum	Calculates sum of elements together	63
98	tan	Tangent for angle in radians	21
99	tand	Tangent for angle in degrees	21
100	tanh	Hyperbolic tangent	21
101	transpose	Transposes elements of an array	46
102	trapz	Numerical integration with the trapezoidal rule	234
103	ver	Displays versions of the MATLAB and toolboxes	27
104	while	Repeats execution of command/s	86
105	who	Displays variables stored in the Workspace	29
106	whos	Displays Workspace variables and additional information about the variables	29
107	zeros	Creates an array with zero	58

Table A.5 Alphabetical list of commands for plotting and drawing.

No	Command	Description	Page/s
1	axis	Controls axis scaling and appearance	131
2	bar	Draws a vertical bars on the plot	168
3	barh	Draws a horizontal bars on the plot	173
4	bar3	Generates 3D vertical bars on the plot	174
5	box on/off	Draws/removes box on the axes	174
6	clabel	Labels iso-level lines	175
7	close	Closes one or more Figure Windows	122
8	colormap	Sets colors	151
9	contour	Creates a 2D-contour plot	174
10	contour3	Creates a 3D-contour plot	175
11	cylinder	Generates a 3D plot with cylinder	175
12	errorbar	Creates a plot with error bounded points	159
13	figure	Creates the Figure window.	175
14	fplot	Creates a 2D plot of a function	176
15	grid on/off	Adds/removes grid lines	131
16	gtext	Adds a text with the help of the mouse	134
17	help graph2d	Displays list of 2D graph commands	173
18	help graph3d	Displays list of 3D graph commands	173
19	help specgraph	Displays list of specialized graph commands	173
20	hist	Plots a histogram	165
21	hold on/off	Keeps current graph open/close	125
22	legend	Adds a legend to the plot	135
23	loglog	Generates a 2D plot with log axes	176
24	mesh	Creates a 3D plot with meshed surface	146
25	meshgrid	Creates X,Y matrices for further plotting	145
26	pdeModeler	Opens an interface for solving PDEs	377
27	pderect	Draws a rectangle or square	395
28	pdeellip	Draws an ellipse or circle	396

Continued

Table A.5 Alphabetical list of commands for plotting and drawing—cont'd

No	Command	Description	Page/s
29	pdecirc	Draws a circle	396
30	pdepoly	Draws a plygon	397
31	pie	Creates a 2D pie plot	176
32	pie3	Creates a 3D pie plot	176
33	plot	Creates a 2D line plot	112
34	plot3	Creates a 3D plot with points/lines	141
35	plottools	Opens the plot editing tools	140
36	polar	Generates a plot in polar coordinates	177
37	rotate3d on/off	Interactively rotates/stops rotation of a 3D plot	157
38	semilogx	Creates a 2D plot with log-scaled x-axis	161
39	semilogy	Creates a 2D plot with log-scaled y-axis	161
40	sphere	Generates a sphere plot	177
41	stairs	Creates star-like plot	178
42	stem	Creates a 2D stem plot	174
43	stem3	Creates a 3D stem plot	174
44	subplot	Places multiple plots on the same page	127
45	surf	Creates a 3D surface plot	148
46	surfc	Generates surface and counter plots together	178
47	text	Adds a text to the plot	134
48	title	Adds a caption to the plot	133
49	view	Specifies a viewpoint for 3D graph	152
50	tri2grid	Converts solution from triangular to rectangular grid	394
51	xlabel	Adds a label to x-axis	133
52	ylabel	Adds a label to y-axis	134
53	yyaxis left/right	Activates left/right y-axes of the plot	163
54	zlabel	Adds a label to z-axis	142

Index

Note: Page numbers followed by *f* indicate figures and *t* indicate tables.

A

abs, 11–13*t*, 260–263*t*
acos, 11–13*t*, 260–263*t*
acosd, 11–13*t*, 260–263*t*
acot, 11–13*t*, 260–263*t*
acotd, 11–13*t*, 260–263*t*
alloy
   binary, phase transition temperature, 116
   dynamic viscosity, 197
   isobaric heat capacity, 198
   thermal conductivity, 187
   liquid, vapor pressure, 198
AND (&), logical, 49, 260*t*
ans, 10, 17–18, 260*t*
answers, to selected questions and exercises, 64–65, 118–121, 166–167, 199–204, 256–258
arithmetical operations, 11–13
   with arrays, matrices, 29
Arrhenius equation, 116
array, element-wise operations
   division, 33
     exponentiation, 33
     multiplication, 33
arrow, key, 10
asin, 11–13*t*, 260–263*t*
asind, 11–13*t*, 260–263*t*
assignment operator (=), 10, 259–260*t*
atan, 11–13*t*, 260–263*t*
atand, 11–13*t*, 260–263*t*
axis, 78–80, 263–264*t*

B

bar, 100, 263–264*t*
barh, 103–106*t*, 263–264*t*
bar3, 103–106*t*, 263–264*t*
Basic Fitting
   interface, 176–182
   panels, 177–180
   resulting plot, example, 180–182
   save, 179
Black's equation, 46
Boltzmann constant, 18–19, 46, 132, 200
Bose-Einstein, energy distribution, 117
boundary conditions, PDE
   1D, 241–246
     presentation in standard form, 215–216
   2D, 246–251
     Diriclet, 224, 226*f*, 241, 246, 252, 255
     Neumann, 221, 241–243, 249, 252, 255
box on/off, 103–106*t*, 263–264*t*
Boyle temperature, 164
Bravais lattice, volume, 21
Brusselator-type, PDE, 243–246
Buckingham equation, modified, 136
bulk modulus, 57–58

C

ceil, 11–13*t*, 260–263*t*
clc, 10, 260–263*t*
clabel, 103–106*t*, 220–221, 263–264*t*
clear, 18, 260–263*t*
Code Analyzer, 125
colon (:), 24, 259–260*t*
color
   specifiers, 72–73*t*, 85, 101*f*
   triplet (RGB), 90
colormap, 90, 263–264*t*
cooling, radiative, 253
Command Window, 9
comma/s, 81, 259–260*t*
comment (%), 124–125, 130, 259–260*t*
compound
   dielectric constant, 170
   viscosity, 193–195
compressibility, 159–161
   factor, 26, 42, 164–165
     via virial coefficient, fitting, 144–149

265

concentration
　molar, 58–59
　substance, 162, 218, 241, 249
conditional statements
　if ... end, 52, 52t, 260–263t
　if...else...end, 52, 52t, 260–263t
　if ... elseif ... else ... end, 52, 52t, 260–263t
　while ... end, 53–54, 53t, 260–263t
conductivity, 41
　thermal, 152–154, 187–189, 232
　　fitting, 187–189
contour, 103–106t, 263–264t
contour3, 103–106t, 263–264t
convdensity, 15, 15f, 260–263t
conversion
　character, fprintf, 21
　triangular to rectangular, grid, 21
convertor
　hardness, Brinell to Vickers and Rockwell, 150
　pressure, psi to MPa, 127
　temperature, Farenheit to Celsius and Kelvin, 128–129
　viscosity, reyn to Pa.sec and vice versa, 150–152
cos, 11–13t, 13, 260–263t
cosd, 11–13t, 13, 260–263t
cosh, 11–13t, 260–263t
cot, 11–13t, 260–263t
cotd, 11–13t, 260–263t
coth, 11–13t, 260–263t
cross, 37–39t, 260–263t
cross-cap, 117
Current Folder
　about, 125–126
　toolbar, 9
　window, 9
curve fitting, 169–176
　multi-variate (two or more variables), 182–184
cylinder, 103–106t, 263–264t

D

Data Statistics, tool, 100–102
density
　benzene, 115
　current, 46
　displaying, 19–20, 50–51
　engine oil, 97
　liquid, linear fit, 55–57
　metals, screening, 50–51
　solid, 154–155
derivative, calculation, 140–142
Desktop, windows, 7–9
det, 37–39t, 260–263t
dew point, temperature, 162
diag, 37–39t, 260–263t
dielectric constant, 134–135, 170
　fitting, equations, 172
　　with Basic Fitting, 180–182
diff, 141, 260–263t
differential equations
　ordinary, 205–214
　partial, 215–221
　　toolbox, 221–232
diffusion
　1D equation, 241–243
　　Brusselator PDEs, 243–246
　　Neuman boundaries, 249
　　reaction, PDE set, 254
　　spherical coordinates, 254
　2D equation, 249–251
　　coordinate dependent initial conditions, 249–251
diffusivity, thermal, 196, 198, 246, 254
disp, 19, 21–22, 40–42, 260–263t
display, output formats, 18
dissociation, rate, 163
distance, between
　two molecules, 61
　inter-particle (or interatomic, or intermolecular), 109, 136, 163
doc, 16, 260–263t
dot, 37–39t, 260–263t
drawing, PDE Modeler, 232
　buttons, 233–235t
　circle, 233–235t
　complicate geometry shapes, 233–235t
　ellipse, 233–235t

polygon, 233–235t
rectangle, 233–235t
square, 233–235t

E
Editor
  plot tools, 78–83, 263–264t
  window
    Live, 225f
    script, 207
elastic (or elasticity)
  modulus, 23–24
    geopolymer, 39
    material, crack-tip
      stresses, fields, 145
element-wise (or element-by-element)
    operations, 33
ellipsoid, 118
ellipse, plot coordinates, 77
else, 52, 52t, 260–263t
elseif, 52, 52t, 260–263t
end, 25–26, 52–53, 52–53t, 148, 260–263t
errorbar, 94–95, 263–264t
enter, key, 10
enthalpy, silver, 68
entropy
  propane vapor, 74, 75f
  substance, 139–140
eq (==), 47, 260t
equal, elemetwise (==), 47, 260t
equation/s
  nonlinear algebraic, 134–136
  set of linear, 31–32
erf, 11–13t, 260–263t
exp, 11–13t, 260–263t
expansion, thermal, coefficient, 141–142
extrapolation, 133–134
eye, 31, 260–263t
extremal point, finding, 136–138

F
factorial, 11–13t, 260–263t
false, value, 47–49
Figure, window, 68
figure, 90, 103–106t, 263–264t
find, 49, 260–263t
finite elements method (FEM), about, 221
fitting
  polynomial, 169–171
  non-polynomial functions, 172–174
  via optimization, 182–184
fix, 11–13t, 260–263t
floor, 11–13t, 260–263t
flow control, 46–59
fminbnd, 136, 260–263t
fminsearch, 182–184, 260–263t
for, 53, 260–263t
force–elongation, data fit, 197
format, 260–263t
  short, 14–15, 18
  shortE, 18–19
  shortG, 14–15, 18–19, 171
  long, 10, 14–15, 18–19
  longE, 18–19, 60
formatting
  iteractive, plot, 84
  text string, 82–83
  2D plots, 78–83
  3D plots, 90–93
fplot, 52t, 263–264t
fprintf, 19–20, 41–42, 163, 165, 184, 194, 260–263t
friction, coefficient, 22
fugacity, fluidic substance, 165
function
  button, 10
  elementary, math, 13
  extremal point, 136–138
  user-defined, 128–132
    arguments, input/output, 129
    body, 130–131
    definition line, 129–130
    file, 131
    help, line/s, 153–154
    live, creating from m-file, 148
function, 129, 260–263t
fzero, 134–136, 154–155, 165, 260–263t

G
gamma, 11–13t, 260–263t
ge (>=), 47, 260t

global, 130–131, 260–263*t*
goodness, fitting, 174–176
Greek letters, 81, 83*t*
grid, on/off, 78, 263–264*t*
greater
　than (>), 47, 260*t*
　　or equal (>=), 47, 260*t*
gt (>), 47, 260*t*
gtext, 81, 263–264*t*

H
Hall-Petch, equation, 197
hardness number
　Brinell, 21–22, 150
　　conversion, 150
　Vickers, Rockwell, 150
heat (or heat transfer)
　capacity (c_p or c_v)
　　Basic Fitting, 192–193
　　data integration, 140–142, 157
　　low-temperature, expression, 62
　　polynomial fitting, 196
　　silver, 61
　1D equation
　　steady-state (or stationary), 232
　　transient, temperature-dependent
　　　material property, 246
　2D equation
　　steady-state (or stationary), 222–223
　　transient, temperature-dependent
　　　material property, 246–248
help, 14–16, 130, 260–263*t*
help, browser/window, 14–16
help graph2d, 103–107, 263–264*t*
help graph3d, 103–107, 263–264*t*
help specgraph, 103–107, 263–264*t*
hist, 98–99, 263–264*t*
histogram, 98
history, MATLAB®, 1–2
hold on/off, 76*f*, 263–264*t*

I
ideal gas law
　3D plot, 88, 88–89*f*
　projections, 93*f*

i, imaginary part of complex number, 17–18, 260*t*
initial conditions
　ODE, 208
　PDE, 1D, 217–218
　　coordinate-dependent, 243–244
　PDE Modeler, 228
　　coordinate-dependent, 249–251
identity matrix, 31
if, 113–114, 260–263*t*
indefinite loop, 54
inf, 17–18, 260*t*
input, function parameters(or arguments), 129
input, 127–128, 260–263*t*
integer, random number, 36
integral, 138–140, 260–263*t*
integration, method
　Simpson, 138
　trapezoidal, 139
intensity
　color, 90
　light, 198
　stress
　　factor, 22
　　in Live Editor, 145
interp1, 133–134, 260–263*t*
interpolation, 133–134
　method
　　cubic (or PCHIP), 134
　　spline, 134
　　linear, 134
inv, 32, 260–263*t*
inverse
　matrix, 32
　trigonometric functions, 13

J
j, imaginary part of complex number, 17–18, 260*t*

K
Kinematic viscosity, motor oil, 115
kinetics, reaction rate, 116
　polyesterification, 240

L

Langmuir, adsorption, 117
lattice
  vector magnitude, 61
  volume, 21
    via vectors, 42–43
  unit vector, 61
launching, MATLAB®, 7–22
ldivide, elementwise (.\), 259–260t
left division (\), 11–13, 32, 259–260t
legend, 171, 179, 263–264t
length, 37–39t, 260–263t
Lennard–Jones, interatomic potential, 109–110
le (<=), 260t
less than (<), 47, 260t
  or equal (<=), 47, 260t
light, intensity, in solid, 198
line specifiers, 69–71t
linewidth, property, 72–73t
Live
  Editor, 144–149
    functions, 148
    scripts, 145–148
linspace, 24, 186, 188, 260–263t
loops, 53–55
  interrupt, ctr and c simultaneously, 54–55
log, 11–13t, 260–263t
log10, 11–13t, 260–263t
logarithm, base a, 11–13t
logical, operators (commands), 48–51, 260t
loglog, 173, 263–264t
lookfor, 130, 149, 260–263t
lt (<), 260t

M

magnitude, vector, 61
marker, type specifiers, 69–71t
markeredgecolor, property, 72–73t
markerfacecolor, property, 72–73t
markersize, property, 72–73t
matrix
  addition/substruction, 28–29
  division, 31–32
  identity, 31
  inverse, 32
  multiplication, 29–31
  size, 37–39t
max, 37–39t, 260–263t
Maxwell-Boltzmann, equation, 113
mean, 37–39t, 260–263t
mean time to failure, MTTF, 45–46
median, 37–39t, 260–263t
mesh, 86–88, 220–221, 230–231, 263–264t
meshgrid, 87, 110, 263–264t
m-file, 3, 148
  opening as live script, 149
M-Lint, messages, 124f
Mie-Grüneisen, equation of state, 154
min, 37–39t, 260–263t
minus (-), 259–260t
mldivide (\), 259–260t
mlx-file, 148
Möbius, strip, 117
modifiers, for text, 82–83
mpower (^), 259–260t
mrdivide (/), 259–260t
mtimes (*), 259–260t

N

NaN, 17–18, 260t
ne (~=), elementwise, 260t
norm of residuals, R$_{norm}$, 177, 192
not (~), logical, 47, 259–260t
not equal to (~=), 47, 259–260t
number
  display formats, 18
  maximal, positive real, 54–55
num2str, 37–39t, 50, 197, 260–263t
numerical methods
  ODE solving, 206–207, 214–215t
  PDE, 1D, solving, 216–217
  PDE, 2D, finite elements, 221–232

O

ode15s, 207–213, 260–263t
ode23, 214–215t, 260–263t
ode23s, 214–215t, 260–263t
ode23t, 214–215t, 260–263t
ode23tb, 214–215t, 260–263t
ode45, 207–213, 240, 260–263t

Index

ode113, 214–215t, 260–263t
ODE solver
  forms, command, 207–213
    extended forms, 212–213
    reducing ODE order, 205–214
    solution methods, numerical, 216–217
    solution steps, 209–212
odeset, 212, 260–263t
ones, 35, 260–263t
or (|), logical, 49, 260t
output
  commands, 19–21
  formats, types, 18–19
  parameters/arguments of function, 129

P

pchip, 143–144t, 260–263t
pdecirc, 233–235t, 263–264t
PDE
  solver, 1D, 215–216
    compliance to standard form, 216
    standard equation form, 215–216
    solution steps, 222–229
  Modeler, 2D, 221
    export, solution and mesh, 230–231
    interface, 221
    solution steps, 222–229
    types of PDEs, 221
pdeellip, 233–235t, 263–264t
pdeModeler, 221–229, 232, 236–238t, 246–251, 263–264t
pdepe, 217–218, 260–263t
pdepoly, 233–235t, 263–264t
percent, symbol (%), 10, 67, 259–260t
permeability, electromagnetic, fitting, 199
pderect, 233–235t, 263–264t
pi, number, 11–13t, 260t
pie, 103–106t, 263–264t
pie3, 103–106t, 263–264t
plate, temperature
  rectangular, 62, 222–223, 246, 263–264t
  square, 111–112
Plank's
  constant, 18
  law, radiance, 155–157

plot (or graph)
  axis limits, 78–80
  bar
    two- or three-dimensional, 100, 103–107
    horizontal, 103–106t
  color, line specifiers, 68
  contour
    2D (or two-dimensional), 103–106t
    3D (or three-dimensional), 103–106t
  editor, 78–83
  formatting, plot editor, 78–84, 90–93
  generation
    XY (2D), 67
    XYZ (3D), 85
  grid, 78
  histogram, 98–99
  labels, 80–81
  legend, 82–83
  mesh
    presenting, 3D, 86–88
    PDE Modeler, 227–228
  multiple
    curves (two or more) in a plot, 74–76
    plots on a page, 76–78
    viewing (or observation angles), 90–91
  projections, 90–93
  properties (line or marker or color), specifiers, 69–71t
  Property Inspector, 84
  rotation, 93
  semi-logarithmic axes, 95–96
  specialized, 94–97
  statistical, 98–102
  stress-strain, 108–109
  text, 81
  two y-axes, 96–97
  title, 80–81
plot, 67, 263–264t
plot3, 84–85, 263–264t
plottools, 84, 263–264t
plus (+), 69–71t, 259–260t
Poisson PDE, 255
polar, 103–106t, 263–264t
polyder, 143–144t, 260–263t
polyfit, 170, 172–173, 260–263t

polymeric materials, kinetics, 240
polyval, 170, 260–263t
power (.^), elementwise, 33, 259–260t
presentation, order, 4–5
Prandtl, number, 198
property, plot
 name, 68
 value, 68
purpose, book, 2–3

Q

quad, 138–139, 260–263t
quadl, 143–144t, 260–263t
quadgk, 143–144t, 260–263t
quad2d, 143–144t, 260–263t

R

radioactive
 decay, 62, 165
 material, amount, 253
rand, 35, 260–263t
randi, 37–39t, 260–263t
randn, 35, 260–263t
random numbers, 35
rate
 dissociation, 163
 reaction, 116, 240
rdivide (./), elementwise, 259–260t
reaction-diffusion, two-equation system, 253
Redlich-Kwong
 compressibility factor, 165
 equation, 164
regression, fitting, 169
relational operators/commands, 48, 260t
repmat, 37–39t, 260–263t
resistivity, metallic conductor, 163
reshape, 37–39t, 260–263t
RGB, triplet, 90
right division (/), 11–13, 32, 259–260t
rng, 36, 260–263t
roots, 143–144t, 260–263t
rotate3d on/off, 93, 263–264t
round, 11–13t, 260–263t
R-squared, criterion, 174

S

Schrodinger-type equation, time-independent, 253
script
 creating file, 123–126
 input values, via Command Window, 127–128
 naming, 125
 running, 126
 saving, 123–126
 live, creating, 145–148
search, Help window, documentation box/field, 9, 16
second order ODE, unforced, 209–212
self-testing, 4, 59–64, 114–118, 161–165, 195–199, 251–256
semicolon (;), 10, 41, 209, 259–260t
semi-logarithmic axis, 95–96
semilogx, 95–96, 263–264t
semilogy, 95–96, 263–264t
sin, 11–13t, 260–263t
sind, 11–13t, 260–263t
sinh, 11–13t, 260–263t
Simpson's rule, 138, 260–263t
size, 37–39t, 260–263t
solution, molar concentration, 58
sort, 37–39t, 260–263t
sphere, 103–106t, 263–264t
spline, 143–144t, 260–263t
sqrt, 11–13t, 260–263t
stairs, 103–106t, 263–264t
statistics
 tool, Data Statistic, molecular weight, 100–102
 calculations, diffusion coefficient, 44–45
std, 37–39t, 260–263t
stem, 103–106t, 263–264t
stem3, 103–106t, 263–264t
stopping indefinite loop, 75–76
strength-grain diameter, data fitting, 197
stress
 field, near crack-tip, 145
 in contact, 60
 intencity factor, 22, 145
  with controls, 149

stress *(Continued)*
  strain, plot, 108–109
    data fitting, 189–191
strvcat, 37–39t, 260–263t
string
  inputing, 40
  output with
    disp, 19, 41–42, 260–263t
    fprintf, 41–42, 260–263t
    input, 127–128, 260–263t
    text, 81, 263–264t
structure
  datatype, 182
  chapters, 4
subplot, 76–77, 79–80, 263–264t
subscript, in text string, 83t
sum, 37–39t, 260–263t
superscript, in text string, 82–83
surf, 86, 89, 263–264t
surface
  coverage, 117
  tension, 107–108, 196
surfc, 103–106t, 263–264t

T
table, displaying, 41–42
tan, 11–13t, 260–263t
tand, 11–13t, 260–263t
tanh, 11–13t, 260–263t
Taylor series, 55
tension, surface, 107–108, 196
text, 81, 263–264t
text, string, 81
  formatting, 82–83
  modifiers, 83t
thermal pressure coefficient, 157–159
times (.*), 259–260t
title, 80–81, 263–264t
toolboxes, about, 16–17
toolstrip
  Desktop, 8–9
  Editor window, 123
  Live Editor, window, 144
topics, about, 3–4
toroid, 117

transpose ('), operator, 28, 59, 260–263t
trapezoidal method (or rule), integration, 139
trapz, 139–140, 260–263t
tri2grid, 231, 255, 263–264t
true, value, 48–49

U
unit vector, 61
user-defined
  function, creation, 128–131
  file, creation, 131

V
vapor pressure-temperature, fitting, 198
variable
  about, 17–18
  global or local, 130–131
  name, 17
  predefined, 17–18, 260t
  management, 17–18
van der Waals
  equation of state, 116, 159, 183–184
  fitting by optimization, P,v,T, 182–184
velocity, gas molecule, 113–114
vector
  creating, 23–25
  matrix operations, 28–32
viscoelastic-type ODE, 209
viscosity
  data fitting, 193–195, 197
  liguid, propane, 63
  motor oil, 96
  units, conversion, 150–152
ver, 16, 260–263t
virial coefficient, fitting, 164
version, about, 4
view, 90–93, 263–264t
view, angles, 90–93

W
wave equation, 255
weight
  histogram, 98–99
  polymer chain, 98
while, 53t, 260–263t

who, 18, 260–263t
whos, 18, 260–263t
Workspace window, 9

X
xlabel, 80–81, 263–264t

Y
yield, strength, 197
ylabel, 80–81, 263–264t

Young's modulus, 18, 63
yyaxis left/right, 96–97, 263–264t

Z
zeros, 35, 260–263t
zlabel, 85, 263–264t